Lecture Notes in Economics and Mathematical Systems

368

Mario Fedrizzi Janusz Kacprzyk
Marc Roubens (Eds.)

Interactive Fuzzy Optimization

Springer-Verlag
Berlin Heidelberg New York London Paris
Tokyo Hong Kong Barcelona Budapest

Managing Editors
Prof. Dr. M. Beckmann
Brown University
Providence, RI 02912, USA

Prof. Dr. W. Krelle
Institut für Gesellschafts- und Wirtschaftswissenschaften
der Universität Bonn
Adenauerallee 24–42, D-5300 Bonn, FRG

Editors
Prof. Mario Fedrizzi
Institute of Computer Science
University of Trento
Via Inama 13, 38100 Trento, Italy

Prof. Janusz Kacprzyk
Systems Research Institute
Polish Academy of Sciences
ul. Newelska 6, 01-447 Warsaw, Poland

Prof. Marc Roubens
Institute of Mathematics
University of Liège
Avenue des Tilleuls (D1)
4000 Liège, Belgium

ISBN 3-540-54577-8 Springer-Verlag Berlin Heidelberg New York
ISBN 0-387-54577-8 Springer-Verlag New York Berlin Heidelberg

Typesetting: Camera ready by author
Printing and binding: Druckhaus Beltz, Hemsbach/Bergstr.
42/3140-543210 – Printed on acid-free paper

PREFACE

The title of this book seems to indicate that the volume is dedicated to a very specialized and narrow area, i.e., to the relationship between a very special type of optimization and mathematical programming. The contrary is however true.

Optimization is certainly a very old and classical area which is of high concern to many disciplines. Engineering as well as management, politics as well as medicine, artificial intelligence as well as operations research, and many other fields are in one way or another concerned with optimization of designs, decisions, structures, procedures, or information processes. It is therefore not surprising that optimization has not grown in a homogeneous way in one discipline either. Traditionally, there was a distinct difference between optimization in engineering, optimization in management, and optimization as it was treated in mathematical sciences. However, for the last decades all these fields have to an increasing degree interacted and contributed to the area of optimization or decision making. In some respects, new disciplines such as artificial intelligence, descriptive decision theory, or modern operations research have facilitated, or even made possible the interaction between the different classical disciplines because they provided bridges and links between areas which had been developing and applied quite independently before.

The development of optimization over the last decades can best be appreciated when looking at the traditional model of optimization.

For a well-structured, i.e., mathematically well-modelable problem, the optimal solution or decision was looked for. This was normally done in a deterministic, certain environment and using a single criterion for evaluating different available or feasible alternatives, The procedure or method of optimization was generally a batch-type process, assuming a closed model in which all necessary information was available and in which the decision maker could provide and process all information simultaneously. Modern optimization theory has relaxed almost all of these limiting assumptions. It had been realized that the use of a single criterion for the evaluation of alternatives was insufficient and unrealistic. Hence, optimization theory moved from single criterion optimization to a huge area of multi criteria analysis. Already in the beginning of the 1970's, people started to realize that human decision makers were not in a position to either provide or process all relevant information and, therefore, one moved from a batch-type methodology to interactive procedures in which only local-type information had to be acquired from the decision maker, in which the simultaneous search procedure was decomposed and transformed into an interactive dialog between a decision maker and powerful data processing systems. Finally – and this is probably the newest development – people did not only realize that a deterministic setting for problems is quite inadequate, but they also accepted that even the use of probability theory to handle uncertainty in problem structures was insufficient. This insight led to the development of theories and methods, such as possibility theory, evidence theory, belief theory and fuzzy sets theory, which do not substitute probability theory but which complement probability theory in the modeling of various kinds of uncertainty. They also facilitate the translation of natural language description of problems into formal models to which powerful mathematical methods can be applied. As usual, the generalization of a rather limited theory leads to richer structures, to more sophisticated models, but also to tougher problems to be solved. The present volume tries to cover al facets of the above mentioned generalizations of a classical optimization model and is, therefore, extremely comprehensive and general in its scope.

The book is an edited volume and is not written by a single author, and I think there are good reasons for this.

Realizing the scope of the book, it is unlikely that a single author could cover in an appropriate way all facets of the problem. This concerns the contributions to the theory as well as the application of the available tools. Therefore, it seems by far advantageous to pool the efforts of specialists in their areas to the benefit of the readers. In this book this approach has been administered particularly well because the more specialized contributions have been linked to each other by broad and introductory coverage of the three major components of this book: Interactive Methods of Multi Criteria Decision Making, Fuzzy Set Theory, and Fuzzy Optimization and Mathematical Programming. One might ask why has optimization been related in particular to mathematical programming?. I believe that this is also a very natural relationship because mathematical programming has been regarded for decades as one of the major application areas of optimization. The methods which have been developed in this area are probably the most frequently applied mathematical methods in practice.

Apart from the introductory papers, which I mentioned above, the book covers the intended area in three major steps:

The second part of the book covers primarily the acquisition of information in uncertain, i.e., fuzzy environments. the second key part of the book focuses on multi criteria decision making and mathematical programming. Predominantly, mathematical programming models or problems in this part of the book are no longer treated in the traditional sense but already as fuzzy mathematical programming models, i.e., as optimization models with constraints in which uncertainty of the problem to be solved has already been integrated into the mathematical model. Nowadays, this type of mathematical programming has already been covered in the literature extensively, and the contributions in the book complement the existing literature in an attractive way.

The last part of the book focuses more or less on interactive decision making and is therefore closets to real applications. It is natural that the contributions consider fuzzy and multi objective problems, and it seems also very timely that all contributions concern computer-related expert or decision support systems.

Altogether, this book represents a very comprehensive treatment of decision making with several criteria, and in uncertain and ill-structured environments. It is especially useful since it also includes introductory survey articles which facilitate the understanding of the more specialized contributions for the newcomer to this area. The book is particularly valuable and recommendable, however, for researchers and practitioners who are competent and have already worked in at least one of the areas of applications or theoretical development, and who want to either update or expand his knowledge in neighboring areas. I hope that this book will become a standard reference in the area of optimization around the world.

Aachen, January 1991 Hans-Jürgen Zimmermann

Table of Contents

1. INTRODUCTORY SECTIONS

FUZZY SET THEORY AND MODELLING OF NATURAL LANGUAGE SEMANTICS

Vilém Novák
Czechoslovak Academy of Sciences, Mining Institute,
Studentská 1768, 708 00 Ostrava-Poruba, Czechoslovakia

Abstract: The paper is a survey of main notions and results obtained in fuzzy set theory with a special emphasise on its use in the modelling of natural language semantics. First, the concept of a fuzzy set and the main notions and operations related to fuzzy sets are presented and a connection of this theory with fuzzy logic is mentioned. Then, a brief overview of results in the modelling of natural language semantics using fuzzy sets is given.
Keywords: Fuzzy sets, Fuzzy logic, Residuated lattice, Natural language semantics.

1 INTRODUCTION

Vagueness is something which makes the human being possible to record complex and relevant phenomena and to disregard unimportant, marginal ones. As expressions of a natural language reflect the way of human recording of phenomena in the world, vagueness becomes one of the most outstanding features of the semantics of natural language. On the other hand, vagueness is one of the forms of indefiniteness. The other form of indefiniteness is uncertainty. This is present when an event is to occur, while vagueness is encountered when we want to record a phenomenon, no matter whether its occuring makes sense on not. For example, it makes sense to speak about the (notion of) tree but we may not have any special tree on our mind and thus, no event is in under the consideration. Therefore, probability theory which may be considered to be a mathematical theory of uncertainty, should not be confused with fuzzy set theory (which is sometimes the case). One of the theories enabling us to grasp the phenomenon of vagueness in a formal way is fuzzy set theory. There are many successful applications of its in various fields and now it is becoming to be successful also in industry, especially in Japan. The core of these applications lays in the possibility to take such factors into account which are describable only by means of natural language. As fuzzy set theory can be used for the modelling of semantics of simple expressions of natural language, we have a tool at our disposal by means of which we can use also vaguely described factors which cannot be handled by classical mathematics.

This paper is aimed at providing the reader with an introductory material which could serve as a basis for a deeper study of fuzzy set theory. The next section is an overview of basic notions and operations. Section 3 contains a brief presentation of the main ideas of the modelling of natural language semantics using fuzzy sets.

2 FUZZY SET THEORY

2.1 The concept of a fuzzy set

Set theory of any kind stems from the assumption that objects can be grouped together. The resulting groupings of objects can be viewed as new objects which, however, can be determined unsharply, i.e. they can be vague. The problem of unsharp determination of a grouping viewed as an object, say A, can be reduced to the question whether a given object, say x, belongs to it. In symbols, we ask whether

$$x \in A \tag{1}$$

(i.e. x is an element of A) is true. If the grouping A is vague then we are not able in general to answer "yes" or "no". Thus, we may seek some kind of a scale on which we could express a degree of truth of (1). We naturally come to many-valued logic which is usually called *fuzzy logic.* The use of a scale for characterization of an unsharp grouping is quite natural for the human mind and it brings us to what can be called a *fuzzy approach.*

Let a basic grouping of objects in which we want to separate a new one be given. The former grouping can be considered to be given sharply, i.e. it may be a set. The newly separated grouping can be unsharp. A classical example is, e.g., the problem of how to separate a grouping of heights of people which fit a vague phenomenon of *being tall.* Then the basic set is, e.g. the interval of real numbers $\langle 40, 300\rangle$ (in cm) as the heights of any human being belongs to it. Note that there are heights belonging to this interval which certainly do not meet the property of "being tall" as well as those which certainly do. Any element from the above mentioned basic set is assigned an element from the scale measuring the extent to which we understand the former element to be a member of the unsharp grouping. More formally, let U be a basic set of elements and L a scale with the greatest element $\mathbf{1}$ and the smallest one $\mathbf{0}$. Considering the above example, the height of 50 cm should be assigned the element $\mathbf{0}$, while that of 200 cm the element $\mathbf{1}$. However, there are also heights which are doubtful, e.g. 172 cm. Does this height mean "tall" or not?

We summarize the above considerations as foollows. Let A be an unsharp grouping of objects taken from U. Then each element of A is characterised by an element of L, i.e. we write

$$A = <x, c>, x \in U, c \in L. \tag{2}$$

The element $c \in L$ is understood to be a *degree of truth* of the proposition "x is an element of A". Since the assignement of the elements $c \in L$ in (2) is unique, A is a function

$$A : U \longrightarrow L. \tag{3}$$

This function is called a *fuzzy set.* The element $A(x) \in L$ is usually written without brackets, i.e. as Ax and it is called the *degree of membership* of x in A. If $Ax = \mathbf{1}$ then x belongs to A. If $Ax = \mathbf{0}$ then x does not belong to A.

Many authors (also in this volume) use the symbol μ (sometimes also f_A, χ_A , etc.) for the function (3) and call it the *membership function* of the fuzzy set A. The symbol A is then used only as a symbol for the unsharp grouping taken as an object. In our notation, the fuzzy set is identified with the membership function which seems to be more transparent and typographically simpler.

If A is a fuzzy set in U then we write $A \underset{\sim}{\subseteq} U$. We will explicitly write a fuzzy set $A \underset{\sim}{\subseteq} U$ in the form

$$\{Ax/x; x \in U\}$$

or

$$A = \bigcup_{x \in U} Ax/x \tag{4}$$

where $Ax \in L$ is a degree of membership of the element $x \in U$. The fuzzy set $\{Ax/x\}$ is called a *fuzzy singleton*. The form (4) in fact means that A is the union of fuzzy singletons. The set of all the fuzzy sets in U is denoted by $\mathcal{F}(U)$.

One may see that the scale of truth values plays a crucial role in fuzzy set theory. There are many studies devoted to intuitive and formal requirements on its structure (see e.g. Bellman and Giertz (1973), Negoita and Ralescu (1975), Termini et al. (1984), Yager (1982), and others). In general, it must be a complete lattice. There are good reasons to assume it to be a *residuated lattice*, i.e. a complete, infinitely distributive lattice

$$\mathcal{L} = \langle L, \vee, \wedge, \otimes, \rightarrow, \mathbf{0}, \mathbf{1} \rangle \tag{5}$$

where $\mathbf{0}, \mathbf{1}$ are the smallest and the greatest elements, $\vee$ and $\wedge$ are the operations of *supremum* and *infimum*, $\otimes$ is the operation of *bold multiplication* (product) and $\rightarrow$ is the operation of *residuation*, respectively, with the following properties:

(a) $\langle L, \otimes, \mathbf{1} \rangle$ is a commutative monoid.

(b) The operation $\otimes$ is isotone in both variables and $\rightarrow$ is antitone in the first variable and isotone in the second one.

(c) The adjunction property

$$a \otimes b < c \text{ iff } a < b \rightarrow c$$

holds for every $a, b, c \in L$.

The reasons for accepting such a structure are analysed in Pavelka (1979), Novàk and Nekola (1983), Novák and Pedrycz (1988) and Novák (1989a). As a special case, most often used in applications, the degrees of truth are assumed to be based on the interval of real numbers $L = \langle 0, 1 \rangle$. Then we put $\vee$:=max, $\wedge$:=min,

$$a \otimes b = 0 \wedge (a + b - 1)$$

and

$$a \rightarrow b = 1 \wedge (1 - a + b)$$

for all $a, b \in \langle 0, 1 \rangle$. The use of this interval is very natural but it is sometimes a cause of confusion between the truth values and probabilities. Let us stress that this is only a formal similarity as the same sets of measures have been used for the two, essentially different phenomena.

In the next subsection we will present some fundamental notions and operations on fuzzy sets.

2.2 Main notions and operations with fuzzy sets

The concepts of fuzzy set theory presented in the sequel, together with many others, can be found in an extensive literature. See e.g. Negoita and Ralescu (1975), Dubois and Prade (1980), Novák (1989a), Zadeh (1983) and others.

A *support* of a fuzzy set A is a classical set

$$\mathrm{Supp}(A) = \{x; Ax \neq \mathbf{0}\}.$$

Similarly, a *kernel* of a fuzzy set A is a classical set

$$\mathrm{Ker}(A) = \{x; Ax = \mathbf{1}\}.$$

A fuzzy set A is *normal* if $\mathrm{Ker}(A) \neq \emptyset$.

A fuzzy set $A \underset{\sim}{\subset} U$ is a *fuzzy subset* of $B \underset{\sim}{\subset} U, A \subseteq B$, if

$$Ax \leq Bx$$

holds for every $x \in U$.

The a–cut of A is a classical set

$$A = \{x; Ax > a\}$$

where $a \in L$.

The following typology of fuzzy sets is often used in applications: Let $\preceq$ be a linear ordering in the universe U and $A \underset{\sim}{\subset} U$ be a fuzzy set. Then A is an S^+–fuzzy set if

$$x \preceq y \text{ implies } Ax < Ay,$$

and it is an S^-–fuzzy set if

$$x \preceq y \text{ implies } Ax > Ay$$

for every $x, y \in U$. If there is a point $y_0 \in U$ such that

$$Ax < Ay_0 \text{ and } Az < Ay_0$$

holds for every $x \preceq y_0$ and $y_0 \preceq z$ then A is a Π–fuzzy set. One of possible general formulae for all the three kinds of fuzzy sets is the following:

$$F(x, a_1, b_1, c_1, c_2, b_2, a_2) = \begin{cases} 0 & \text{if } x < a_1 \text{ or } x > a_2 \\ 1 & \text{if } c_1 \leq x \leq c_2 \\ \frac{1}{2}\left(\frac{x-a_1}{b_1-a_1}\right)^2 & \text{if } a_1 \leq x < b_1 \\ 1 - \frac{1}{2}\left(\frac{x-c_1}{c_1-b_1}\right)^2 & \text{if } b_1 \leq x < c_1 \\ 1 - \frac{1}{2}\left(\frac{x-c_2}{b_2-c_2}\right)^2 & \text{if } c_2 < x \leq b_2 \\ \frac{1}{2}\left(\frac{x-a_2}{a_2-b_2}\right)^2 & \text{if } b_2 < x \leq a_2 \end{cases} \tag{6}$$

where $a_1, a_2 \in U, a_1 < a_2$ are left and right edge points, $c_1, c_2 \in U, c_1 \leq c_2$ are the edge points of the kernel and $c_1, c_2 \in U$ are points usch that $Ab_i = 0.5, i = 1, 2$. When setting $a_1 = b_1 = c_1$, we obtain the S^- curve and $a_2 = b_2 = c_2$ yields the S^+ one.

Let $A, B \underset{\sim}{\subset} U$. The basic operations with fuzzy sets are:

union

$$C = A \cup B \text{ iff } Cx = Ax \vee Bx$$

intersection

$$C = A \cap B \text{ iff } Cx = Ax \wedge Bx$$

bold intersection

$$C = A \boxtimes B \text{ iff } Cx = Ax \otimes Bx$$

residuum

$$C = A \,\textcircled{\scriptsize$\rightarrow$}\, B \text{ iff } Cx = Ax \rightarrow Bx.$$

On the basis of residuum, one can define the *complement*

$$\overline{A} = A \,\textcircled{\scriptsize$\rightarrow$}\, \emptyset.$$

where $\emptyset$ is the empty fuzzy set $\emptyset = \{0/x; x \in U\}$. Note that if $L = \langle 0, 1\rangle$ then this definition gives $\overline{A}x = 1 - Ax$ for all $x \in U$, i.e. we obtain the original definition of the complement given in Zadeh (1965).

A specific feature of fuzzy sets is the possibility to define also new, additional operations. Let us stress that this possibility is very important in the applications of fuzzy sets. First, we introduce two new operations of
biresiduation

$$a \leftrightarrow b = (a \rightarrow b) \wedge (b \rightarrow a)$$

and the *p-th power*

$$a^p = \underbrace{a \otimes \ldots \otimes a}_{p\text{–times}}$$

for all the $a, b \in L$. Note that

$$a \leftrightarrow b = 1 - \mid a - b \mid$$

holds for $L = \langle 0, 1\rangle$.

Let o be an n–ary operation on L to be added. Then the following *fitting condition* must be fulfilled: there are $p_1, \ldots, p_n$ such that

$$(a_1 \leftrightarrow b_1)^{p_1} \otimes \ldots \otimes (a_n \leftrightarrow b_n)^{p_n} \leq o(a_1, \ldots, a_n) \leftrightarrow o(b_1, \ldots, b_n) \tag{7}$$

holds for every $a_i, b_i \in L, i = 1, \ldots, n$. The justification of the fitting condition can be found in Pavelka (1979) and Novák and Pedrycz (1988). Note that all the basic operations fulfil the fitting condition. Moreover, the folowing holds true:

Theorem 1 *All the the operations derived from the operations fulfilling the fitting condition fulfil it, as well.*

This theorem makes us sure that the class of fitting operations is quite rich. Thus, we are able to enrich the lattice $\mathcal{L}$ by new, additional operations $o : L^n \longrightarrow L$. Using them, we may define new n–ary operations with fuzzy sets as follows. Let and $A_1, \ldots, A_n \underset{\sim}{\subset} U$ be fuzzy sets. Then o is a basis of the operation O assigning a fuzzy set $C \underset{\sim}{\subset} U$ to $A_1, \ldots, A_n$ when we put

$$C = O(A_1, \ldots, A_n) \text{ iff } Cx = o(A_1 x_1, \ldots, A_n x_n) \tag{8}$$

for every $x \in U$.

It is easy to prove:

Theorem 2 *Let $\{o_d; d \in I\}$ be a set of fitting operations enriching the lattice $\mathcal{L}$ which are the bases of the corresponding operations $\{O_d; d \in I\}$ due to (8). Then*

$$\langle \mathcal{F}(U), \cup, \cap, \otimes, \ominus, \emptyset, U, \{O_d; d \in I\}\rangle$$

is a residuated lattice enriched by the operations $\{O_d; d \in I\}$.

This theorem gives rise to the definition of new operations on fuzzy sets and demonstrates their properties. Some examples of the additional operations with fuzzy sets are the following ($L = \langle 0, 1\rangle$):
algebraic product

$$C = A \cdot B \text{ iff } Cx = Ax \cdot Bx$$

algebraic sum

$$C = A \oplus B \text{ iff } Cx = Ax + Bx - Ax.Bx$$

bounded difference

$$C = A \ominus B \text{ iff } Cx = 0 \vee (Ax - Bx),$$

and many others.

Let U be a linearly ordered set. A fuzzy set $A \underset{\sim}{\subseteq} U$ is *convex* if

$$Ax_1 > Ax_2 \wedge Ax_3$$

holds for every $x_1, x_2, x_3 \in U, x_1 < x_2 < x_3$.

A very important notion is that of a fuzzy *cardinality* of a fuzzy set. There are several kinds of them (Zadeh, 1983, Novák 1989a). If Supp$(A), A \underset{\sim}{\subseteq} U$ is finite then we may define:
Absolute fuzzy cardinality

$$\text{FCard}(A) = \{\alpha_n/n; n \in N\} \tag{9}$$

where

$$\alpha_n = \bigvee\{\beta; \text{Card}(A_\beta) = n\}$$

and A_β is a β–cut of A.
Relative fuzzy cardinality of A with respect to B, where $A, B \underset{\sim}{\subseteq} U$:

$$\text{FCard}_A(B) = \{\alpha_r/r; r \in Re\} \tag{10}$$

where

$$\alpha_r = \bigvee\{\beta; \frac{\text{Card}(A_\beta \cap B_\beta)}{\text{Card}(A_\beta)} = r\}$$

Other important notions of fuzzy set theory are connected with the notion of the Cartesian product. Let $U = U_1, \times \ldots \times U_n$ and $A_i \underset{\sim}{\subseteq} U_i, i = 1, \ldots, n$. The *Cartesian product* of fuzzy sets $A_1, \ldots, A_n$ is a fuzzy set

$$(A_1, \ldots, A_n) < x_1, \ldots, x_n >= A_1x_1 \wedge \ldots \wedge A_nx_n.$$

A fuzzy set

$$R \underset{\sim}{\subseteq} U_1 \times \ldots \times U_n$$

is called a *fuzzy relation.* Let $R \underset{\sim}{\subseteq} U \times V, S \underset{\sim}{\subseteq} V \times W$ be fuzzy relations. Then the composition of fuzzy relations is a fuzzy relation

$$T = R \circ S \underset{\sim}{\subseteq} U \times W$$

defined by

$$(R \circ S) < x, z >= \bigvee_{y \in V} (R < x, y > \wedge S < y, z >). \tag{11}$$

The operation $\wedge$ in (11) can be replaced by $\otimes$. The resulting composition of fuzzy relations is called a *weak composition* and we denote it by $R \overset{\circ}{\times} S$.

There are natural definitions of the *inverse, domain* and *range* of fuzzy relations.

Many papers are devoted also to the problem of solution of fuzzy relational equations: given fuzzy relations R and T, or S and T, the task is to find a relation S so that $R \circ S = T$ $(R \circ S = T)$. This problem was first solved by Sanchez (1976).

A powerful principle of fuzzy set theory is the *extension principle* which makes possible to extend any operation on U to an operation on fuzzy sets.

Let $* : U \times U \longrightarrow U$ be a binary operation on U and $A, B \underset{\sim}{\subset} U$ be fuzzy sets. Then the extension of "$*$" on fuzzy sets A, B is the fuzzy set $C = A * B$ with the membership function

$$Cz = \begin{cases} \bigvee\{Ax \wedge Bx; x, y \in U \text{ and } z = x * y\} \\ \text{o otherwise} \end{cases}$$

for every $z \in U$.

A convex normal fuzzy set $A \underset{\sim}{\subset} Re$ is called a *fuzzy number*. Using the extension principle, all the common operations with fuzzy numbers may be defined. For details see Dubois and Prade (1978, 1980, 1988), Ramík and Římánek (1985), Novák (1989a) and others.

3 THE USE OF FUZZY SETS IN MODELLING OF NATURAL LANGUAGE SEMANTICS

3.1 Introduction to the problem

The theory of fuzzy sets seems to be a suitable tool to be used for the modeling of semantics of natural language (at least of its part). In this section, we briefly present the main ideas concerning this task.

A sentence of natural language can be viewed from several points. In classical linguistics, it is usual to talk about the representation of a sentence on various levels. In the system called the *functional generative description of natural language* (FGD), (see Sgall, Hajičová and Panevová (1986), or Sgall (1984)) five levels are differetiated, namely a *phonetic* (PH) (how a sentence is composed as a system of sounds), a *phonemic* (PM) (how words of a sentence are composed), a *morphemic* (MR) (how a sentence is composed of its words), *surface syntax* (SS) (the system of grammatical rules) and *tectogrammatical* (TR) which is the highest level corresponding to the semantics. The latter is also called a *deep structure* of the sentence and this is an objective of possible application of fuzzy set theory. Viewed on this level, words and more complex syntagms of natural language can be understood to be names of propertics enocountered by man in the world. Fuzzy sets can be used as follows: let $\mathcal{A}$ be a syntagm [1] of natural language and φ the corresponding property in its deep structure. Provided that it leads to a fuzzy set $A \underset{\sim}{\subset} U$, the meaning of $\mathcal{A}$ is set

$$M(\mathcal{A}) = A. \tag{12}$$

Thus, our job consists in determining the membership function A.

However, the situation is by no means simple as not every word of natural language coresponds to such a property and, above all, there are various relations between the words. Thus, the determination of the membership function which corresponds to a complex syntagm may be a very complicated task. Moreover, one surface structure of a sentence may lead to several deep structures. Up to now, we are far from a detailed understanding to all the nuances of the sentential semantics. The present state of the art makes us possible to model the meaning of only some simple syntagms.

3.2 The meaning of some types of syntagms

First, let us stop at the modelling of the semantics of nouns. In general, if S is a noun, then its meaning is a fuzzy set

$$M(\mathcal{S}) = S, \ S \underset{\sim}{\subset} U.$$

[1] A syntagm is any part of the sentence formed according to the grammatical rules

What is the universe U? It is a set of objects chosen in such a way that whenever an object x has property φ_S labeled by the noun S then $x \in U$. This can be constructed, e.g., as follows: Let K be a set of generic elements called the *kernel space.* For example, K can be a union of all the objects described in our dictionary, of those we have regarded during the last week, of those we see in our flat etc. In short, K should contain all the specific objects we have met or may imagine. Let $\mathcal{F}(K)$ be the set of all the fuzzy sets on K and put

$$\mathcal{F}^n(K) = \underbrace{\mathcal{F}(\ldots(\mathcal{F}(K)\ldots)}_{n\text{–times}}$$

Let E_K be the smallest set closed with respect to all the Cartesian powers of K, of $\mathcal{F}^n(K), n = 1, \ldots, n$ and all the Cartesian products of these elements. This set is called the *semantic space.* Then the universe of S is a sufficiently big subset $U \subseteq E_K$.

A certain problem is the determination of the membership function. There are several methods proposed in the literature (cf. Novák, 1989a). The membership function corresponding to object nouns (e.g. *table, car, donkey,* etc.) could be constructed on the basis of outer characteristics of elements. For example, we may use proportions of some geometric patterns contained in objects etc. A very often used method is a statistical analysis of expert (subjective) estimations. Several experiments have been described in the literature. Let us mention that fuzzy methods are rather robust and thus an exact determination of a membership function is not as important as it might seem at the first glance. The experience suggests that even an individual estimation works well when it is done carefully and seriously. For this purpose, we may successfully use the formula (6).

In practical applications, e.g. in artificial intelligence, it is not very useful to model the meaning of nouns because we would have to find a proper representation of its elements which, in fact, we do not need. Most applications are based on modelling of the meaning of adjectives and the syntagms of the form

(quantifier —) adverb — adjective (— noun)

where the syntagm

adverb — adjective (13)

plays a crucial role. The most important (and very frequent) adjectives are those inducing an ordering $\leq$ in the universe U. We will assume that $\leq$ is linear. According to lingustitic considerations as well as experiments (see Kuz'min, 1985), there are certain points $m, s, v, \in U$ where $m < s < v$. The point s is called a *semantic center.* The adjectives inducing an ordering in U usually form antonyms which can be characterized as follows. Let $\mathcal{A}^-, \mathcal{A}^+$ be antonyms (e.g. small – big, cold – hot etc.). Then their meanings are fuzzy sets A^-, A^+ respectively, such that A^- is an S^-–fuzzy set and $\mathrm{Supp}(A^-) \subseteq \langle m, s)$, and A^+ is an S^+–fuzzy set and $\mathrm{Supp}(A^+) \subseteq (s, v\rangle$. We will often call $\mathcal{A}^-$ a *negative* and $\mathcal{A}^+$ a *positive* adjective, respectively. There are also couples of antonyms such that a third member $\mathcal{A}^0$ exist. Its meaning is a fuzzy set A^0 with the property $s \in \mathrm{Ker}(A^0)$. A typical example of the adjective $\mathcal{A}^0$ is $\mathcal{A}^0$:=*average.* In the sequel we will call $\mathcal{A}^0$ a *zero* adjective.

The adverb in (13) is an intensifying one (e.g. *very, highly, absolutely, slightly* etc.) and it is usually called a *linguistic modifier* in fuzzy set theory. In general, the meaning of an intensifying adverb m is a pair of functions

$$M(m) = \langle \zeta_m, \nu_m \rangle$$

where $\zeta_m : U \longrightarrow U$ is a displacement function and $\nu_m : L \longrightarrow L$ is a unary operation fitting L [2]. Hence, the meaning of the syntagm (13) is obtained using the composition of functions

$$M(m\mathcal{A}) = \nu_m \circ A \circ \zeta_m. \tag{14}$$

[2]Some authors simplify this model by putting $\zeta_m = \mathrm{id}_U$ (an identical function on U). However, this is linguistically incorrect.

For some modifiers, the function ν_m has been experimentally derived as follows:

$$\begin{aligned}
\nu_{very}(a) &= a^2 \\
\nu_{more_or_less}(a) &= 2a - a^2 \\
\nu_{highly}(a) &= a^3 \\
\nu_{roughly}(a) &= -a^4 + 4a^3 - 6a^2 + 4a
\end{aligned}$$

for all $a \in \langle 0,1\rangle$. The estimation of function ζ_m is more difficult. For example, the modifier *very* may yield the following:

$$\zeta_{very}(x) = x + (-1)^k \cdot d \cdot \parallel \mathrm{Ker}(A) \parallel$$

where $k = 1$ for A^+ or for A^0 if $x \le s$, and $k = 2$ for A^- or for A^0 if $x \ge s$. $\parallel \mathrm{Ker}(A) \parallel$ is the length of the interval $\langle \inf(\mathrm{Ker}(A)), \sup(\mathrm{Ker}(A))\rangle$. The parameter d was experimentally estimated to be a number $d \in \langle 0.25, 0.40\rangle$.

Examples of some other linguistic modifiers can be found in the literature.

The meaning of verbs is a very complicated problem and so far, only the copula "to be" in syntagms as

$$p := \mathcal{P} \text{ is } \mathcal{A} \tag{15}$$

is modelled where $\mathcal{A}$ is usually the syntagm (13). However, (15) is interpreted rather as a simple assignement than a verb. The $\mathcal{P}$ in (15) is a noun but it is usually not treated as such. Thus, we obtain two ways of how the meaning of (15) can be modelled.

a) We put

$$M(p) = M(\mathcal{A}) = A,$$

i.e. the meaning of p is set equal to the meaning of syntagm $\mathcal{A}$. This is quite reasonable since, as was stated above, we usually need not know the meaning of noun $\mathcal{P}$ in the applications.

b) Let $M(\mathcal{P}) = P \underset{\sim}{\subseteq} V$. Then we put

$$M(p) \subseteq P \times A. \tag{16}$$

The inclusion in (16) may be proper or improper depending on the kind of noun $\mathcal{P}$. The relation (16) means that each element x from the universe V (a representative of noun $\mathcal{P}$) is asigned an attribute y from the universe U [3] where $A \underset{\sim}{\subseteq} U$ [4].

A conditional sentences of the form

$$\mathcal{C} := \text{IF } p \text{ THEN } q \tag{17}$$

where p and q are syntagms of the form (15) often occur in applications. In fuzzy set theory we usually put

$$M(\mathcal{C}) = M(p) \ominus M(q), \tag{18}$$

i.e. we interpret the implication (17) using the residuum operation between fuzzy sets $M(p)$ and $M(q)$.

Let us remark that many authors interpret (17) as the Cartesian product

$$M(\mathcal{C}) = M(p) \times M(q). \tag{19}$$

---

[3] U is usually the real line. For example, in the syntagms such as *Peter is tall* , the element *Peter* is assigned its height being a real number.

[4] Zadeh and some other authors following him (also in this volume) suggest to interpret the membership degree $(P \times A) < x, y >$ as a *possibility degree* of the fact that x is A. However, the possibility degree concerns uncertainty which, in our oppinion, does not reflect the vagueness phenomenon contained in the semantics of natural language.

In applications, this may work since all fuzzy methods are very robust and the implication (17) often describes only some kind of a relation between the input and output, i.e. it is not, in fact, understood to be the implication. However, putting the meaning of the implication (17) equal to the Cartesian product is linguistically as well as logically incorrect since (19) is symmetric and the implication is not.

Let us also mention the problem of linguistic quantifiers (we will denote them by the letter $\mathcal{Q}$). They do not form a uniform group form the linguistic point of view. We place among them numerals including the indefinite ones (e.g. *several*), some adverbs (e.g. *many, few, most*), some pronouns (e.g. *every*), some nouns (e.g. *majority, minority*) and others. From the point of fuzzy set theory, their meaning is generally a fuzzy number. In Zadeh (1983), an interpretation of syntagms of the form

$$\mathcal{Q}\mathcal{A} \tag{20}$$

or

$$\mathcal{Q}\mathcal{A}'s\ are\ \mathcal{B}'s \tag{21}$$

is proposed. In (20), quantifier $\mathcal{Q}$ is interpreted as a fuzzy characterization of the absolute fuzzy cardinality (9) of the fuzzy set

$$A = M(\mathcal{A})$$

while in (21), it characterizes the relative fuzzy cardinality (10) of $M(\mathcal{A})$ with respect to $B = M(\mathcal{B})$. More exactly, we put

$$M(\mathcal{Q}\mathcal{A}) = Q \cap \mathrm{FCard}(A) \tag{22}$$

and

$$M(\mathcal{Q}\mathcal{A}'s\ are\ \mathcal{B}'s) = Q \cap \mathrm{FCard}_A(B). \tag{23}$$

However, the problem is not finished yet since (22) and (23) make sense only for fuzzy sets with finite supports.

In the literature on fuzzy set theory (see, e.g., Zadeh, 1973, 1975, 1978b, 1983, Novák, 1989a, and others), one may find the semantics of the compound syntagms of the form

$$\mathcal{A}\ and\ \mathcal{B} \tag{24}$$

and

$$\mathcal{A}\ or\ \mathcal{B} \tag{25}$$

defined using the operations of intersection and union of fuzzy sets, respectively. However, this is only a tentative solution since the syntagms (24) and (25) are special cases of a very complicated phenomenon known in linguistics as the *coordination.* The use of the operations of intersection and union of fuzzy sets may work in some special cases of the close coordination in such syntagms as, e.g., *Peter and Paul ..., Old and dirty car ...*, etc.

An even worse situation is encountered with the negation. A simple use of the operation of complement of fuzzy sets works only with some kinds of adjectives and nouns. However, the negation contained in more complex syntagms is incidental to the phenomenon of the *topic–focus articulation* when only the focus is being negated. A fully comprehensive description of this phnomenon in linguistics is not, however, yet done.

A very important concept which deserves to be mentioned here is that of a *linguistic variable* (Zadeh, 1975). This concept makes it possible to see a certain part of linguistics from a more technical point of view.

A linguistic variable is in general a quintuple

$$\langle \mathcal{X}, \mathcal{T}(\mathcal{X}), U, G, M \rangle$$

where $\mathcal{X}$ is a name of the variable, $\mathcal{T}(\mathcal{X})$ is its term-set, U is a universe, G is a syntactic and M a semantic rule, respectively. For example,

$$\mathcal{X} := size,$$

$$U := \langle 0, 1000 \rangle,$$

G is a certain, usually context-free grammar generating the set $\mathcal{T}(\mathcal{X})$ of terms such as *big, small, very big, rather average* etc. and M is a syntactic rule assigning to each term $\mathcal{A} \in \mathcal{T}(\mathcal{X})$ its meaning being a fuzzy set

$$M(\mathcal{A}) \underset{\sim}{\subseteq} U.$$

Lingustic variables play an important role in applications. For example, parameters of a technical system such as *temperature, speed, weight* etc. can be understood to be lingustic variables. As its values can also be crisp, e.g. *exactly 165.2* etc., the concept of a linguistic variable is general enough to capture also the classical concept of a variable.

3.3 Linguistic approximation

In applications, one may also meet the problem of a *linguistic aproximation.* We may lay it down as follows. Let $\mathcal{T}$ be a set of syntagms of natural language, e.g., the term set of a linguistic variable and let us be given a fuzzy set $A_0' \underset{\sim}{\subseteq} U$. Our task is to find a syntagm $\mathcal{A}_0 \in \mathcal{T}$ such that its meaning $M(\mathcal{A}_0) = A_0$ is as close to A_0' as possible.

There are many ways how to solve this task. However, no sufficiently efficient and general method is known till now.

If $\mathcal{T}$ has a small number of elements, then we may succesfully use the following method. Find $\mathcal{A}_0$ such that

$$\sum_{x \in U} \mid A_0 x - A_0' x \mid^p \tag{26}$$

is minimal for some suitable, previsously set number p. This method is often used in technical applications.

A quite effective procedure was proposed in Esragh and Mamdani (1979). This procedure is suitable for syntagms of the form (24) and (25) where $\mathcal{A}$ and $\mathcal{B}$ may consist of an adjective, a noun and a linguistic modifier, and the universe U is ordered. According to this method, the membership function A_0' is divided into parts by effective turning points (i.e. special points where the membership function changes its course), the parts are approximated by the above partial syntagms $\mathcal{A}, \mathcal{B}$ using (26) and the resulting syntagm is obtained by joining $\mathcal{A}$ and $\mathcal{B}$ using the corresponding connective. In particular, if the two neighbouring parts of the membership function form a "hill", then the corresponding syntagms are joined by the connective *and*, and if they form a "valley", then they are joined by the connective *or*. Note that this works only in the case when the connective *and* is interpreted as the intersection and *or* as the union of fuzzy sets.

4 CONCLUSION – PERSPECTIVES OF FUTURE DEVELOPMENTS

In this paper, we have briefly presented main ideas of fuzzy set theory with emphasise on its use in the modelling of natural language semantics. Concerning the latter, we attempted to demonstrate how the semantics of some of the basic units can be interpreted, namely the semantics of nouns, adjectives, selected adverbs and the copula "to be". Moreover, the semantics of some cases of the close coordination (the use of connectives) was also touched along with the semantics of conditional sentences.

Fuzzy set theory is becomming more and more popular and very many books and papers are available. But only few of them deal with the theory of the semantics of natural language though many of them apply the theory presented above. Therefore, it is necessary to elaborate it more deeply in cooperation with the linguists. Let us stress that we are still far from grasping of the meaning of even quite simple sentences when they contain a verb. The reason consists in an extreme complexity of verb semantics since verbs represent the most important units of our language stepping towards the human's recording of the surrounding world at the highest level of his intellectual capability. But there are also some questions concerning the very basis of the notion of a fuzzy set which must still be answered. Some work on both matters is done in Novák (1989b) where, however, a new world of mathematics called the *alternative set theory* (AST) is used (see Vopěnka, 1979). In the light of this model, fuzzy set theory serves there as a special technical tool which can be used at a second stage after the semantics of a sentence (syntagm) in the frame of AST is formed.

Despite the above facts, the use of fuzzy sets in the modelling of natural language semantics has already found many successful applications, mainly in Japan where fuzzy sets are used in many commercially available products. This is a convincing argument in favour of the usefullness and sense of fuzzy set theory.

REFERENCES

Bezdek, J.,ed. (1987a) **Analysis of Fuzzy Information - Vol. 1: Mathematics and Logic**, CRC Press, Boca Raton, Fl.

Bezdek, J., ed. (1987b) **Analysis of Fuzzy Information - Vol. 2: Artificial Intelligence and Decision Systems**, CRC Press, Boca Raton, Fl.

Bezdek, J., ed. (1987c) **Analysis of Fuzzy Information - Vol. 3: Applications in Engineering and Science**, CRC Press, Boca Raton, Fl.

Dubois D., and Prade H. (1980) **Fuzzy Sets and Systems:Theory and Applications**, Academic Press, New York.

Dubois D., and Prade H. (1988) **Possibility Theory - An Approach to Computerized Processing of Uncertainty**. Plenum Press, New York.

Esragh, F., Mamdani, E.H. (1979) *A general approach to linguistic approximation*, Int. J. Man-Mach. Stud., **11**, 501-519.

Gupta, M.M., Yamakawa, T., eds. (1988) **Fuzzy Computing: Theory, Hardware and Applications**, North–Holland, Amsterdam.

Gupta, M.M., Yamakawa, T., eds. (1988) **Fuzzy Logic in Knowledge–Based Systems, Decision and Control**, North–Holland, Amsterdam.

Kuz'min, V. B. (1985) *About semantical structure of linguistic hedges:an experimental hypothesis*, BUSEFAL **24**, 118 - 125, Université Paul Sabatier, Toulouse.

Lakoff, G. (1973) *Hedges: A study in meaning criteria and logic of fuzzy concepts*, J. Philos. Logic **2**, 458 - 508.

Negoita C. V., and Ralescu D. A. (1975) **Application of Fuzzy Sets to System Analysis**. Birkhauser, Stuttgart.

Novák V. (1989a) **Fuzzy Sets and Their Applications**. Adam Hilger, Bristol.

Novák, V. (1989b) **The Alternative Mathematical Model of Natural Language Semantics.** Manuscript. Mining Institute, Ostrava 1989. (To be published by Cambridge University Press)

Novák V., and Pedrycz W. (1988) *Fuzzy sets and t-norms in the light of fuzzy logic.* Int. J. Man-Machine Stud. 29, 113-127.

Pavelka J. (1979) *On fuzzy logic I, II, III.* Zeit. Math. Logik Gurndl. Math. **25**, 45-52, 119-134, 447-464.

Sanchez E. (1976) *Resolution of composite fuzzy relation equations.* Inf. Control 30, 38-48.

Sgall, P., ed. (1984) **Contributions to functional syntax, semantics, and language comprehension**, Academia, Prague.

Sgall, P., Hajičová, E., Panevová (1986) J., **The meaning of the sentence in its semantic and pragmatic aspects**, D. Reidel, Dordecht.

Skala H. J., Termini S., and Trillas E., eds. (1984) **Aspects of Vaguenes**. Theory and Decision vol 39. D. Reidel, Dordecht.

Vopěnka P. (1979) **Mathematics In the Alternative Set Theory**. Teubner, Leipzig.

Yager R. R., ed. (1982) **Fuzzy Set and Possibility Theory. Recent Developments**. Pergamon Press, New York.

Zadeh L. A. (1965) *Fuzzy sets.* Inf. Control 8, 338-353.

Zadeh, L.A. (1973) *Quantitative Fuzzy Semantics*, Inf.Sci.,**3**, 159-176.

Zadeh L. A. (1975) *The concept of a linguistic variable and its application to approximate reasoning I, II, III.* Inf. Sci. 8, 199-257, 301-357; 9, 43-80.

Zadeh L. A. (1978a) *Fuzzy sets as a basis for a theory of possibility.* Fuzzy Sets and Systems 1, 3-28.

Zadeh L. A. (1978b) *PRUF - a meaning representation language for natural languages.* Int. J. Man-Machine Stud. 10, 395–460.

Zadeh, L.A. (1983) *A computational approach to fuzzy quantifiers in natural languages*, Comp. Math. with Applic. **9**, 149-184.

A SURVEY OF FUZZY OPTIMIZATION AND MATHEMATICAL PROGRAMMING

Mario Fedrizzi*, Janusz Kacprzyk**
and José L. Verdegay***

* Institute of Computer Science, University of Trento, Via Rosmini 42, 38100 Trento, Italy
** Systems Research Institute, Polish Academy of Sciences, ul. Newelska 6, 01-447 Warsaw, Poland
*** Department of Computer Science and Artificial Intelligence, University of Granada
18071 Granada, Spain

Abstract
A brief survey of various concepts, problem classes, issues, etc. related to fuzzy optimization and fuzzy mathematical programming is provided. Emphasis is on various approaches to fuzzy linear programming as the most important from both the practical point of view and the purpose and scope fo the volume.

Keywords
Fuzzy Optimization, Fuzzy Mathematical Programming, Fuzzy Linear Programming.

1. INTRODUCTION

Optimization is viewed here as a kind of decision making in which information of the preferences among the alternatives is described by some utility (objective, performance, ...) function that maps a set of feasible alternatives into $\mathbb{R}$. On the other hand, *mathematical programming* is viewed as a kind of optimization when the set of feasible alternatives is described by a system of equations and/or inequalities.

Optimization (mathematical programming) has been successfully used for years in a variety of problems related to «hard» systems (e.g., technological) in which the structure, relations, behavior, etc. are well-defined and quantifiable. Unfortunately, attempts to apply similar means to «soft» systems (social, economic, ...) have not been in general successful. A key role is there played by human judgments, preferences, etc. which are subjective, imprecise and not easily quantifiable.

A promising tool for «softening» optimization and mathematical programming models seems to be Zadeh's *fuzzy sets theory* which has recently experienced an unprecedented growth of interest.

In this survey our basic philosophy is to replace fuzzy mathematical programming problems by their equivalent (in terms of an optimal solution) nonfuzzy counterparts. We concentrate, first, on some general concepts, issues and problem classes of fuzzy optimization and then on fuzzy linear programming.

Our notation related to fuzzy sets is standard, and we will informally equate fuzzy sets with their membership functions.

For other earlier surveys, see above all Kacprzyk and Orlovski (1987b) which is close to this survey as to its perspective, scope, etc.; Luhandjula's (1989) review is more limited as to the scope, and from a slightly different point of view. The reader is also referred (in addition to the source papers) to the articles in Kacprzyk and Orlovski (1987a).

2. APPROACHES TO FUZZY OPTIMIZATION WITH AN EXPLICITLY SPECIFIED FEASIBLE SET

An optimization problem involves two essential elements: (1) a *set of feasible alternatives* (*a feasible set*), and (2) an *objective* (performance, ...) *function* used for the comparison of alternatives; then, an optimal alternative is sought.

The class of problems considered here may be generally stated as follows. $X = \{x\}$ is a set of alternatives. The objective function is $F: X \to L(\mathbb{R})$ where $L(\mathbb{R})$ is a class of fuzzy subsets of $\mathbb{R}$; i.e. $F(x)$ is a fuzzy number being an imprecise evaluation of $x \in \mathbf{X}$. The feasible set is described by a fuzzy set $C \subset \mathbf{X}$ whose membership function is $\mu_C: \mathbf{X} \to [0,1]$; $\mu_C(x) \in [0,1]$ is the degree of feasibility of x.

A fuzzy optimization problem is now generally stated as

$$F(x) \to \underset{x \in C}{\widetilde{\mathrm{Max}}} \tag{1}$$

where «$x \in C$» denotes «x 'belongs' to C», and « $\widetilde{\max}$ » denotes the maximization in a fuzzy sense, (1) schould be read as to attain a 'possibly high' fuzzy value of F with respect to x's 'belonging' to C.

2.1. *Attainment of a fuzzy goal subject to a fuzzy constraint. Bellman and Zadeh's approach*

In Bellman and Zadeh's (1970) approach, we have an explicit fuzzy feasible set, $C \subset \mathbf{X}$, called a *fuzzy constraint*, and an explicit fuzzy set of alternatives that attain a goal, $G \subset \mathbf{X}$, called a *fuzzy goal*.

The value of $\mu_G(x)$ indicates the degree to which $x \in \mathbf{X}$ satisfies G. For example (see, e.g., Kacprzyk, 1983a)

$$\mu_G(x) = \begin{cases} 1 & \text{for} \quad f(x) \geq \bar{f} \\ g(x) & \text{for } \underline{f} \leq f(x) \geq \underline{f} \\ 0 & \text{for} \quad f(x) \leq \underline{f} \end{cases} \tag{2}$$

to be read as: we are fully satisied ($\mu_G(x) = 1$) with the x's for which f(x) attains a value not below an aspiration level $\bar{f}$; we are less satisfied (to degree $0 < g(x) < 1$) with the x's for which f(x) is between $\bar{f}$ and a lowest acceptable level $\underline{f}$; finally, we are fully dissatisfied with the x's for which f(x) is below $\underline{f}$. And similarly for C.

This aspiration-level-based interpretation of G (and C) is not the only possible though is certainly very convenient and intuitively appealing.

The problem is now stated as to «satisfy C and attain G» which by introducing a fuzzy set $D \subset \mathbf{X}$ to be a solution and, called a *fuzzy decision*, can be written as

$$\mu_D(x) = \mu_C(x) \wedge \mu_G(x) = \min(\mu_C(x), \ \mu_G(x)), \forall x \in \mathbf{X} \tag{3}$$

Though «$\wedge$» is the most commonly used operation, also here, other operations, notably t-norms, may be used.

A next problem is which $x \in \mathbf{X}$ is to be chosen as a (nonfuzzy) solution to (3). As commonly assumed, we seek an $x^* \in \mathbf{X}$ such that

$$\mu_D(x^*) = \sup_{x \in \mathbf{X}} \mu_D(x) \tag{4}$$

For multiple fuzzy constraints, $C_1, \ldots, C_m \subset \mathbf{X}$, and multiple fuzzy goals, $G_1, \ldots, G_m, \subset \mathbf{X}$, (3) becomes

$$\mu_D(x) = \mu_{C_1}(x) \wedge \ldots \wedge \mu_{C_m}(x) \wedge \mu_{G_1}(x) \wedge \ldots \wedge \mu_{G_m}(x), \forall x \in \mathbf{X} \tag{5}$$

and an $x^* \in \mathbf{X}$ is sought such that (4) holds.

Moreover, if we have $C \subset \mathbf{X}$, $G \subset \mathbf{Y}$ and a function $w: \mathbf{X} \to \mathbf{Y}$, $y = w(x)$, then (3) becomes

$$\mu_D(x) = \mu_C(x) \wedge \mu_{G'}(w(x)), \forall x \in \mathbf{X} \tag{6}$$

where $G' \subset \mathbf{X}$ is a fuzzy goal in $\mathbf{X}$ induced by $G \subset \mathbf{Y}$. Then, evidently, an $x^* \in \mathbf{X}$ is sought via (4).

Suppose that the goal is to attain some fuzzily specified value of the objective function (fuzzy goal) $\mu_G: \mathbb{R} \to [0,1]$. First, consider a precisely defined objective function, i.e. $f: \mathbf{X} \to \mathbb{R}$. The subset of alternatives providing for the satisfaction of G is the inverse image $\bar{\mu}_G$ of μ_G under f, i.e.

$$\bar{\mu}_G(x) = \mu_G(f(x)), \forall x \in \mathbf{X} \tag{7}$$

If we replace $\mu_G(x)$ in (6) by $\bar{\mu}_G(x)$, then

$$\mu_D(x) = \mu_C(x) \wedge \mu_G(f(x)), \forall x \in \mathbf{X} \tag{8}$$

and we seek an $x^* \in \mathbf{X}$ such that (4).

Now, consider a more general case when the values of the objective function are fuzzy, i.e. $f: \mathbf{X} \times \mathbb{R} \to [0,1]$.

Orlovski (1981) introduced a fuzzy decision $D \subset \mathbf{X}$ which is a maximal fuzzy set (with respect to the containment «$\subset$») satisfying:

1) $D \subset C$ (feasibility of the solution)
2) $D \circ f \subset G$ (attainment of the fuzzy goal)

where $D \circ f$ is the image of D under $f : \mathbf{X} \times \mathbb{R} \to [0,1]$.

D is here evidently equivalent to (8) in the case of a precise (nonfuzzy) objective function.

Orlovski (1980) introduces first the following sets:

$$\mathcal{N} = \{(x, r) : (x, r) \in \mathcal{N} \times \mathbb{R}, f(x, r) > \mu_G(r)\} \tag{9}$$

$$\mathcal{N}_x = \{r : (r \in \mathbb{R}, (x, r) \in \mathcal{N}\} \tag{10}$$

$$\mathbf{X}^\circ = \{x : x \in \mathbf{X}, \mathcal{N}_x \neq \emptyset\} \tag{11}$$

Then

$$\mu_D(x) = \begin{cases} \mu_C(x) \wedge \inf\limits_{r \in \mathcal{N}_x} \mu_G(x) & \text{for } x \in \mathbf{X}^\circ \\ \mu_C(x) & \text{otherwise} \end{cases} \tag{12}$$

which for $f: \mathbf{X} \to \mathcal{E}$ boils down to (8).

As previously, we seek an $x^* \in \mathbf{X}$ such that (4).

2.2. *The use of α-cuts of the fuzzy feasible set*

As opposed to goal-satisfaction-type approaches, the following Orlovski's (1977) approach is more explicitly based on the concept of maximization.

The problem is as in (1), i.e. $f(x) \to \widetilde{\mathrm{Max}}_{x \in C}$. To derive the first solution concept we employ the α-cuts of C, $C_\alpha = \{x \in \mathbf{X} : \mu_C(x) \geq \alpha\}$, $\forall \alpha \in (0,1]$. Then, for each α such that $C_\alpha \neq \emptyset$ we introduce the (nonfuzzy) set

$$\mathcal{N}(\alpha) = \{x \in \mathbf{X}: f(x) = \sup_{x \in C_\alpha} f(z)\} \tag{13}$$

By *solution* 1 to (1) we mean a fuzzy set

$$\mu_{S_1}(x) = \begin{cases} \sup\limits_{x \in \mathcal{N}(\alpha)} \alpha & \text{for } x \in \bigcup\limits_{\alpha > 0} \mathcal{N}(\alpha) \\ 0 & \text{otherwise} \end{cases}$$

$$= \begin{cases} \mu_C(x) & \text{for } x \in \bigcup\limits_{\alpha > 0} \mathcal{N}(\alpha) \\ 0 & \text{otherwise} \end{cases} \tag{14}$$

New, we define the fuzzy maximal value of f(x) over C as

$$\mu_f(r) = \sup_{x \in f^{-1}(r)} \mu_{S_1}(x) = \sup_{x \in f^{-1}(r)} \sup_{x \in \mathcal{N}(\alpha)} \alpha, \quad \forall r \in \mathbb{R} \tag{15}$$

Notice that a final solution is not simply an x with the highest $\mu_{S_1}(x)$ but is also based on the value of f (x); if $f(x_o) = r_o$ it is obvious that the greater r_o the smaller $\mu_{S_1}(x_o)$, hence a compromise is needed.

Solution 2 to (1) is based on the concept of the *Pareto maximum*. For f(x) and $\mu_C(x)$ we first define the set of Pareto maximal elements, $\mathcal{P}$, as a (nonfuzzy) subsets $\mathcal{P} \subseteq \mathbf{X}$ such that $x \in \mathcal{P}$ if there exists no $y \in \mathbf{X}$ for which either

$$f(y) > f(x) \text{ and } \mu_C(y) \geq \mu_C(x)$$
$$\text{or} \tag{16}$$
$$f(y) \geq f(x) \text{ and } \mu_C(y) > \mu_C(x)$$

Solution 2 is now defined as a fuzzy set

$$\mu_{S_2}(x) = \begin{cases} \mu_C(x) & \text{for } x \in \mathcal{P} \\ 0 & \text{otherwise} \end{cases} \tag{17}$$

As shown by Orlovski (1977), this solution gives the same fuzzy maximal value of f(x) over C as solution 1, i.e.

$$\mu_f(r) = \sup_{x \in f^{-1}(r)} \mu_{S_2}(x) = \sup_{x \in f^{-1}(r)} \sup_{x \in \mathcal{N}(\alpha)} \alpha, \quad \forall r \in \mathbb{R} \tag{18}$$

Moreover, $\mathcal{P} \subset \bigcup_{\alpha > 0} \mathcal{N}(\alpha)$ which implies $\mu_{S_2}(x) \leq \mu_{S_1}(x)$, $\forall x \in \mathbf{X}$, i.e. solution 2 is a subset of solution 1.

Other approaches based on a separate treatment of f(x) and $\mu_C(x)$ may be exemplified by Negoita and Ralescu (1977), Ralescu (1984) and Yager (1979).

2.3. *The case of multiple objective functions and fuzzy constraints*

With n objective functions and m fuzzy constraints, the problem becomes

$$\{f_1(x), \ldots, f_n(x)\} \rightarrow \overline{\max_{x \in \mathbf{X}}} \qquad \text{subject to: } \mu_{C_1}(x), \ldots, \mu_{C_m}(x) \tag{19}$$

where « $\overline{\max}$ » is the maximization in the sense of Pareto.

Similarly as for a single objective function, the approaches sketched in Subsections 2.1 and 2.2 can be applied. Here we will outline another approach due to Orlovski (1978, 1980, 1984), which is intuitively appealing and efficient.

First, assume that the objective functions take on fuzzy values $F_i(x) \subset \mathbf{X}$ with real-valued functions as special cases. Thus, the membership grade of $r = f_i(x)$ in $F_i(x)$ is $\mu_{F_i(x)}(r)$.

Through $\mu_{F_i}(x)$ we obtain using the extension principle n fuzzy nonstrict preference relations

$$p_i(x_i, x_2) = \sup_{z \geq y} (\mu_{F_i(x_i)}(z) \wedge \mu_{F_i(x_2)}(y)) \tag{20}$$

Next, we define a fuzzy strict preference relation corresponding to p_i as

$$p_i^s(x_i, x_2) = \max\{0, p_i(x_i, x_2) - p_i(x_2, x_1)\} \tag{21}$$

and the degree to which x_1 is strictly preferred to x_2 is

$$p^s(x_i, x_2) = \bigwedge_{i=1}^{n} p_i^s(x_i, x_2) \tag{22}$$

we introduce a fuzzy subset of nondominated alternatives

$$\mu_{ND}(x) = 1 - \sup_{y \in X} p^s(y, x) = 1 - \sup_{y \in X} \bigwedge_{i=1}^{n} p_i^s(y, x) =$$

$$= 1 - \sup_{y \in X} \bigwedge_{i=1}^{n} (p_i(y, x) - p_i(x, y)) \tag{23}$$

where $\mu_{ND}(x)$ is a degree of nondominance of x, i.e. if $\mu_{ND}(x) \geq \alpha$, then x may be strictly dominated by some other alternative to a degree smaller than $1-\alpha$.

Now we define a degree of feasibility of x as

$$\mu_{FS}(x) = \mu_{C_1}(x) \wedge \ldots \wedge \mu_{C_m}(x) \tag{24}$$

The solution of (19) is then to find an $x^* \in \mathbf{X}$ such that

$$\mu_{ND}(x^*) \geq \alpha \text{ and } \mu_{FS}(x^*) \geq \beta \quad (25)$$

where α and β are desidered degrees of nondominance and feasibility (a compromise between α and β is sought).

In Orlovski (1984) some nonfuzzy optimization problems equivalent to (19) are described.

For other approaches based on some idea of nondominance see, e.g., Takeda and Nishida (1980), Leung (1982, 1984) and Carlsson (1982).

3. BRIEF INTRODUCTION TO FUZZY MATHEMATICAL PROGRAMMING

The point of departure is here

$$f(x) \rightarrow \max_{x \in \mathcal{R}^n}$$

$$\text{subject to: } g_i(x) \leq b_i,\ i = 1, \ldots, m \quad (26)$$

where $x = [x_i, \ldots, x_n] \in \mathbb{R}^n$ is a vector of decision variables, f: $\mathbb{R}^n \rightarrow \mathbb{R}$ is an objective fuction, g_i: $\mathbb{R}^n \rightarrow \mathbb{R}$ are constraints, and b_i: $\in \mathbb{R}$ are the right-hand-sides; maximization and «$\leq$» and be easily replaced by minimization and «$\geq$», respectively.

Particularly important in practice is linear programming (LP) meant as

$$\{f(x) = cx = \sum_{j=1}^{n} c_j x_j\} \rightarrow \max_{x_j \geq 0}$$

$$\text{subject to: } (Ax)_i = \sum_{j=1}^{n} a_{ij} x_j \leq b_i,\ i = 1, \ldots m \quad (27)$$

3.1. *Different forms of Fuzzy Linear Programming (FLP) problems*

Looking at (27) we may see the following elements that may be fuzzified. On the one hand, the coefficient (costs) in f (x), $c_1, \ldots, c_n$, the technological m x n matrix $A = [a_{ij}]$, and the right-hand-sides $b = [b_i, \ldots, b_m]$; and on the other hand, the (strict) maximization.

This implies the following main types of FLP problems:

- problems with fuzzy constraints,
- problems with a fuzzy goal (objective function),
- problems with fuzzy costs (c_i' s), and
- problems with fuzzy coefficients (a_{ij}'s and b_i's)

In the FLP of the first type we perfectly know all the parameters but we allow the constraints to be violated up to some extent. The degree of satisfaction of the i-th fuzzy constraint is

$$\mu_i(x) = \begin{cases} 1 & \text{if} \quad (Ax)_i < b_i \\ h_i[(Ax)_i] & \text{if} \quad b_1 \leq (Ax)_i \leq b_i + t_i \\ 0 & \text{if} \quad (Ax)_i > b_i + t_i \end{cases} \quad (28)$$

It is usually assumed that $h_1(.) \in [0,1]$ is such that the higher the violation of the constraints, the lower the value of h_i.

An FLP problem of the second type is when there is a fuzzy goal $\mu_o : \mathbb{R}^n \rightarrow [0,1]$ such that $\mu_o(x) \in [0,1]$ expresses the degree of attainment of a fuzzy goal. Usually, an aspiration level for the objective function is fixed and the fuzzy goal is that the value of the objective function be essentially greater than or equal to that aspiration level. Formally, μ_o is identifical to (28).

In the third type of FLP problems we have in the objective function fuzzy coefficients expressed by fuzzy numbers.

The fourth type of FLP problems is when we have both fuzzy coefficients in the objective function and fuzzy coefficients in the technological matrix.

More complicated cases (combinations of the above four types) will not be discussed here.

3.2. *Methods for solving fuzzy linear programming problems*

3.2.1. The case of a fuzzy constraint set

We consider

$$cx \to \max_{x}$$

$$\text{subject to: } Ax \lesssim b,\ x \geq 0 \tag{29}$$

For each constraint in (29) we have a membership function of type (28), $\mu_i : \mathbb{R}^n \to [0,1]$, $i = 1, \ldots, m$; (29) should be meant as «the left-hand-side should be essentially less that or equal to the right-hand-side» with the understanding that they should be possibly well satisfied.

In the first approach to solving (29) we suppose that the h_i (.)'s in (28) are linear.

Tanaka, Okuda and Asai (1974) seek such an optimal $(\alpha^*, x^*) \in [0,1] \times \mathbb{R}^n$ that

$$\alpha^* \wedge f(x^*) = \operatorname{Sup} \{\alpha \wedge \max_{x \in X_\alpha} f(x)\} \tag{30}$$

where $f: \mathbb{R}^n \to [0,1]$ is a continuous objective function, and $X_\alpha = \{x \in \mathbb{R}^n | \bigwedge_i \mu_i(x) \geq \alpha\}$, $\alpha \in [0,1]$.

Under some mild assumptions concerning the continuity of f and uniqueness of a*, (a*, x*) is obtained using the following algorithm

1) Assume $k = 1$ and $\alpha_k \in [0,1]$.
2) Compute $f_k = \operatorname{Max} \{f(x) | x \in X_{\alpha_k}\}$
3) Compute $\varepsilon_k = \alpha_k - f_k$. If $|\varepsilon_k| > \varepsilon$, go to 4). If $|\varepsilon_k| \leq \varepsilon$, go to 5). ε is some required precision.
4) Let $\alpha_{k+1} = \alpha_k - r_k \varepsilon_k$, replace k by k + 1 and go to 2); $r_k \geq 0$ is selected so that $0 \leq \alpha_{k+1} \leq 1$.
5) Let $\alpha^* = \alpha$ and determine an optimal $x^* \in \mathbb{R}^n$ such that

$$f(x^*) = \operatorname{Max} \{f(x) | x \in X_{\alpha^*}\} \tag{31}$$

On the other hand, Zimmermann's (1976) approach is directly inspired by the concept of a maximizing decision (4). We put $e_j = -c_j$ and consider the following fuzzified version of (29)

$$\begin{aligned} &\sum_{j=1}^{n} e_j x_j \lesssim z \\ &\sum_{j=1}^{n} a_{ij} x_j \lesssim b_i, \quad i = 1, \ldots m \\ &\quad x_j \geq 0,\ j = 1, \ldots, n \end{aligned} \tag{32}$$

meant as follows: ex should be «essentially smaller than or equal to» an aspiration level z, and the constraints should be «essentially smaller than or equal to» b_i; evidently, both should be satisfied as well as possible.

First, to formalize «$\lesssim$» (essentially smaller than or equal to), we denote by $H = [h_{kj}]$ the (m+1) x n–matrix obtained by adding to $A = [a_{ij}]$ the row vector $[e_j]$ before the first row of A. As in the case of A, we denote HX's k-th row by

$$(Hx)_k = \sum_{j=1}^{n} h_{kj} x_j \tag{33}$$

and define the function

$$g_k[(Hx)_k] = \begin{cases} 1 & \text{for} \quad (Hx)_k \leq w_k \\ 1 - [(Hx)_k - w_k]/t_k & \text{for} \quad w_k < (Hx)_k \leq w_k + t_k \\ 0 & \text{for} \quad (Hx)_k \leq w_k + t_k \end{cases} \tag{34}$$

where $w^T = (w_1, \ldots, w_{m+1}) = (z, b_1, \ldots, b_m)$, and t_k's are some admissible violations of the constraints.

We wish to satisfy all the constraints of (32), hence the new objective function is

$$\mu_D(x) = \bigwedge_{k=1}^{m+1} g_k[(Hx)_k] \tag{35}$$

and we seek an optimal $x^* = \{x_1^*, \ldots, x_n^*) \in \mathbb{R}^n$ such that

$$\mu_D(x^*) = \sup_{x \in \mathcal{R}^n} \mu_D(x) \tag{36}$$

It may be shown (Negoita and Ralescu, 1977; Zimmermann, 1976) that each optimal solution of the following LP problem is also an optimal solution of (32)

$$\begin{aligned} &\lambda \rightarrow \max_{\lambda \in [(0,1]} \\ &\text{subject to:} \quad \lambda \leq w'_k - (Hx)'_k \qquad k = 1, \ldots, m+1 \\ &\qquad\qquad\quad x_j \geq 0 \qquad\qquad\qquad\quad j = 1, \ldots, n \end{aligned} \tag{37}$$

where $w'_k = w_k/t_k$ and $(H_k)' = (Hx)_k/t_k$.

The third approach to the solution of (29), proposed by Verdegay (1982), uses the so-called representation theorem (cf. Negoita and Ralescu, 1975) which basically says that a fuzzy set can be uniquely represented by all its α-cuts. The approach may be summarized as follows. If the membership functions of the fuzzy constraints in (29) are strictly monotone and continuous, which is often natural, the α-cuts of the set of constraints can be expressed as

$$C_\alpha = \{x \in \mathbb{R}^n \mid \sum_{j=1}^{n} a_{ij}x_j \leq g_i^{-1}(\alpha);\ x_k \geq 0,\ i=1, \ldots, m;\ j=1, \ldots, m\} \tag{38}$$

where the g_i^{-1} (.)'s are the inverse functions of the $g_i(.)$'s defined in (28). Then, if C denotes the fuzzy constraint set of (29), the representation theorem says that

$$C = \sum_{\alpha \in [0,1]} \alpha C_\alpha \tag{39}$$

Thus, a (fuzzy!) solution to (29) can be obtained by solving the following parametric LP problem

$$\begin{aligned} &cx \rightarrow \max_{x} \\ &\text{subject to: } x \in C_\alpha,\ \alpha \in (0,1] \end{aligned} \tag{39}$$

or more explicitly

$$\begin{aligned} &cx \rightarrow \max_{x} \\ &\text{subject to: } (Ax)_i \leq g_i^{-1}(\alpha),\ \alpha \in (0,1],\ i = 1, \ldots, m \end{aligned} \tag{40}$$

For linear fuzzy constraints, this becomes

$$\begin{aligned} &cx \rightarrow \max_{x \geq 0} \\ &\text{subject to: } Ax \leq b + t\,(1-\alpha),\ \alpha \in (0,1] \end{aligned} \tag{41}$$

where $t = [t_1, \ldots, t_m]$ is a (column) vector representing an allowable violation of each constraint [cf. (28)].

Therefore, if $x^*(\alpha)$ is an optimal solution of (40), from this a fuzzy optimal solution of (29) can be obtained.

Now, to relate (and somehow unify) the above three approaches, denote by f(x) the objective function in (30), and by $f_{m+1}(x)$ the membership function of the fuzzy goal in (34). As shown by Verdegay (1982), the solution of (29), derived by using Tanaka, Okuda and Asai's (1974) algorithm, and the corresponding one of (37), derived by using Zimmermann's (1975) approach, can be obtained from the fuzzy solution of (40), $x^*(\alpha)$, by solving the following equations for a particular $\alpha \in (0,1]$, respectively

$$f[cx^*(\alpha)] = \alpha, \qquad f_{m+1}[cx^*(\alpha)] = \alpha \tag{42}$$

Notice that while solving (29) by using (40) it is not necessary to define f(.) and f_{m+1} (.) a priori. Moreover, the size of (40) is as that of (29).

Finally, if the g_i (.)'s in (38) are not linear, that nonlinarity occurs in the right-hand-sides of (40) but it does not affect the linearity of the auxiliary parametric LP problem; (40) is therefore also valid for nonlinear constraints.

3.2.2. Fuzzy coefficients in the objective function

Now, we have an FLP problem such that the constraints are nonfuzzy and the coefficients in the objective functions, c_j's, are fuzzy numbers $\mu_j: \mathbb{R}^n \rightarrow [0,1]$, j=1, ..., n.

We will now present three approaches to dealing with that kind of problems due to Delgado, Verdegay and Vila (1990), Rommelfanger, Hanuscheck and Wolf (1989) and Tanaka, Ichihashi and Asai (1984).

To start, let us consider an FLP problem with a fuzzy objective in the sense of Verdegay (1984), i.e.

$$\begin{aligned} & cx \rightarrow \max_{x \geq 0} \\ & \text{subject to: } Ax \leq b \end{aligned} \tag{43}$$

with the membership functions of c_j assumed as

$$\mu_j(x) = \begin{cases} 0 & \text{if } \bar{c}_j \leq x \text{ or } x \leq \underline{c}_j \\ \underline{h}_k(x) & \text{if } \underline{c}_j \leq x \leq c_j \\ \bar{h}_k(x) & \text{if } c_j \leq x \leq \bar{c}_j \end{cases} \tag{44}$$

where $[\bar{c}_j, \underline{c}_j]$ is the support of c_j and $\underline{h}_j$ (.) and $\bar{h}_j$(.) are continuous and strictly increasing and decreasing, respectively, functions such that $\underline{h}_j(c_j) = \bar{h}_j(c_j) = 1$.

Then, using a fuzzy objective as defined in Verdegay (1984) and considering the corresponding (1 - α) – cuts of each cost, a∈ [0,1], we have $\forall x \in \mathbb{R}$ and j = 1, ..., n,

$$\mu_j(x) \geq 1 - \alpha \Leftrightarrow \underline{h}_j^{-1}(1-\alpha) \leq x \leq \underline{h}_j^{-1}(1-\alpha) \tag{45}$$

and, denoting $\Phi_j(.)$ and $\psi_j(.) = \underline{h}_j^{-1}(.)$, we obtain $\Phi_j(1-\alpha) \leq x \leq \psi_j(1-\alpha)$.

Then, it can be shown that a fuzzy solution to (43) can be found from the parametric solution of the following multiobjective LP problem

$$\begin{aligned} & [c^1x, c^2x, \ldots, c^{2^n}x] \rightarrow \max_x \\ & \text{subject to: } Ax \leq b,\ x \geq 0 \\ & c^k \in E(1-\alpha),\ \alpha \in [0,1],\ k=1,2,\ldots,2^n \end{aligned} \tag{46}$$

where E(1 - α) is the set of vectors in $\mathbb{R}^n$ each of whose components is either on the upper bound, $\psi_j(1-\alpha)$, or on the lower bound, $\Phi_j(1-\alpha)$, of the respective (1 - α) – cuts, i.e. ∀j = 1, ..., n

$$c = [c_1, c_2, \ldots, c_n] \in E(1-\alpha) \Leftrightarrow c_j = \Phi_j(1-\alpha) \text{ or } \psi_j(1-\alpha) \tag{47}$$

On the other hand, Tanaka Ichihashi and Asai (1984) propose a possibilistic approach, meant for even more general problems with fuzzy constraints. Evidently, it is valid for nonfuzzy constraints as well. This method is based on the two relevant facts. First, that the objective functions's value $z = c_1x_1 + c_2x_2 + \ldots + c_nx_n$, where the c_j's are fuzzy numbers with triangular membership functions, $\Delta(\underline{c}_j, c_j, \bar{c}_j)$, is also a fuzzy number

$$\mu(y) = \begin{cases} 1 - [\,|\,2y - (\bar{c} + \underline{c})\,|\,x] / (\bar{c} - \underline{c})x, & x > 0, y > 0 \\ 1 & x = 0, y > 0 \\ 0 & x = 0, y = 0 \end{cases} \tag{48}$$

where $\bar{c} = [\bar{c}_1, \bar{c}_2, \ldots, \bar{c}_n]$, $\underline{c} = [\underline{c}_1, \underline{c}_2, \ldots, \underline{c}_l]$ and $\underline{c}_j$ and $\bar{c}_j$ are as in (44).

Second, that $z \to \underset{x}{\tilde{\max}}$ is meant as $w_1 \bar{c}x + w_2 cx \to \max_{x}$, where $w_1 + w_2 = 1$, $w_1, w_2 \in [0,1]$ are weights.
Thus a solution of (43) is derived by solving the following auxiliary LP problem

$$
\begin{aligned}
& w_1 \bar{c}x + w_2 \underline{c}x \to \max_{x} \\
& \text{subject to: } Ax \leq b,\ x \geq 0
\end{aligned}
\tag{49}
$$

The solution of (43) is also considered by Rommelfanger, Hanuscheck and Wolf (1989). Their approach is termed a stratified piecewise reduction approach, and is based on Zimmermann's (1975) approach. Though the problem is (43), the vagueness of the coefficients is modeled there by using nested intervals each of which has a membership, or better yet a possibility degree $\alpha_k \in [0,1]$, where $k \in \mathcal{M} = \{1, 2, \ldots, p\}$, i.e. each (vague) coefficient in the objective function is defined by means of

$$c_j = \{[\underline{c}_j, \bar{c}_j]^k/\alpha_k;\ k \in \mathcal{M}\},\ \forall j = 1, \ldots, n \tag{50}$$

such that

$$\forall \alpha_1, \alpha_2, \in [0,1],\ \alpha_1 \geq \alpha_2 \Rightarrow [\underline{c}_j, \bar{c}_j]^1 \subseteq [\underline{c}_j, \bar{c}_j]^2 \tag{51}$$

Evidently, any fuzzy coefficient given as (44) may be formulated as (50) by using its α–cuts.
Then, the solution of (43) is obtained by solving the following auxiliary LP problem (Rommelfanger, Hanuscheck and Wolf, 1989)

$$
\begin{aligned}
& \lambda \to \max_{\lambda \geq 0} \\
& \text{subject to: } f_1[\underline{c}^\alpha x] \geq \lambda,\ f_2[\underline{c}^\alpha x] \geq \lambda\ Ax \leq b,\ x \geq 0
\end{aligned}
\tag{52}
$$

where:

$$f_1[\underline{c}^\alpha x] = (\underline{c}^\alpha x - \bar{z}^\alpha_{min}) / (z^{*\alpha}_{min} - \bar{z}^\alpha_{min}) \qquad \text{if } \bar{z}^\alpha_{min} \leq \underline{c}^\alpha x \leq z^{*\alpha}_{min}$$

$$f_2[\bar{c}^\alpha x] = (\bar{c}^\alpha x - \bar{z}^\alpha_{max}) / (z^{*\alpha}_{max} - \bar{z}^\alpha_{max}) \qquad \text{if } \bar{z}^\alpha_{max} \leq \bar{c}^\alpha x \leq z^{*\alpha}_{max}$$

and if we denote $Z = \{x \in \mathbb{R}^n | Ax \leq b,\ x \geq 0\}$, then:

$$z^{*\alpha}_{min} = \underline{c}^\alpha(x^*_{min}) = \text{Max}\ \{\underline{c}^\alpha x | x \in \mathcal{Z}\};\quad z^{*\alpha}_{max} = \bar{c}^\alpha(x^*_{max}) = \text{Max}\ \{\bar{c}^\alpha x | x \in \mathcal{Z}\}$$

$$\bar{z}^\alpha_{min} = \underline{c}^\alpha(x^*_{max}) = \text{Min}\ \{\underline{c}^\alpha x | x \in \mathcal{Z}\};\quad \bar{z}^\alpha_{max} = \bar{c}^\alpha(x^*_{min}) = \text{Min}\ \{\bar{c}^\alpha x | x \in \mathcal{Z}\}$$

Then the solution of (47) may be found as the insersection of the solutions obtained by solving (52) for each $\alpha_k \in \mathcal{M}$.
Finally, notice that (52) may be rewritten as

$$
\begin{aligned}
& \lambda \to \max_{\lambda \geq 0} \\
& \text{subject to:} \quad Ax \leq b,\ x \geq 0 \\
& \qquad (z^{*\alpha}_{min} - \bar{z}^\alpha_{min})\lambda - \underline{c}^\alpha x \leq \bar{z}^\alpha_{min} \\
& \qquad (z^{*\alpha}_{max} - \bar{z}^\alpha_{max})\lambda - \underline{c}^\alpha x \leq \bar{z}^\alpha_{max}
\end{aligned}
\tag{53}
$$

Then, evidently, $f_1[\underline{c}^\alpha x^*] = f_2[\bar{c}^\alpha x^*] = \lambda^*$ where (λ^*, x^*) is the optimal solution of (53).
The problems (46), (49) and (53) are closely interrelated (see Delgado, Verdegay and Vila, 1989b).

3.2.3. Fuzzy coefficients in the technological matrix

Now the coefficients in the technological matrix and the right-hand-sides are represented by fuzzy numbers; on the other hand, the coefficients in the objective function are nonfuzzy. For simplicity, the fuzzy numbers are of LR type (cf. Dubois and Prade, 1980).
The FLP problem of this type is

$$cx \to \max_{x \geq 0}$$

$$\text{subject to: } \sum_{j=1}^{n} \tilde{a}_{ij} x_j \leq \tilde{b}_i,\ i = 1, \ldots, m \tag{54}$$

This class of problems is discussed by Tanaka, Ichihashi and Asai (1984), even with fuzzy coefficients in the objective function which is not dealt with here.

Since in (54) the fuzziness is in the coefficients, and not in the allowable violation of the constraints, then (29) and (54) are principally different.

For solving (54), Tanaka, Ichihashi and Asai (1984) solve the following auxiliary conventional LP problem

$$cx \to \max_{x \geq 0}$$

$$\text{subject to:}$$
$$[(1-\beta/2)(\underline{a}_i + a_i) + (\beta/2)(a_i - \underline{a}_i)]\, x \leq [(1-\beta/2)(b_i + \underline{b}_i) + (\beta/2)(b_i - \underline{b}_i)$$
$$[(\beta/2)(\bar{a}_i + a_i) + (1-\beta/2)(\bar{a}_i - a_i)]x \leq (\beta/2)(b_i + \bar{b}_i) + (1-\beta/2)(\bar{b}_i - b_i) \tag{55}$$

where $\beta \in [0,1]$ is a degree of optimism to be specified a priori.

To obtain (55), the following ordering relation between the fuzzy numbers is assumed

$$a >^{\beta} b \Leftrightarrow (a + \bar{a})_k \geq (b + \bar{b})_k \text{ and } (a - \underline{a})_k \geq (b - \underline{b})_k \tag{56}$$

for each $k \in [\beta, 1]$, where $(a + \bar{a})_k$ and $(a - \underline{a})_k$ are the upper and lower limit of the k-cut of a, respectively.

Notice that by (56) no fuzziness is in fact assumed as to the satisfaction of the constraints. To account for this, Delgado, Verdegay and Vila's (1989) model of FLP may be employed. It starts from

$$cx \to \max_{x \geq 0}$$

$$\text{subject to: } \sum_{j=1}^{n} \underline{a}_{ij} x_j \lesssim b_i,\ i = 1, \ldots, m \tag{57}$$

However, since the coefficients in the constraints are imprecise, some violation of them may be allowed which, for the i-th constraint, is expressed by a fuzzy number t_i. This can be modeled as follows. The $\mu_i : \mathbb{R} \to [0,1]$ and $\mu_{ij} : \mathbb{R} \to [0,1]$ express the knowledge on b_i and a_{ij}, respectively, and $\mu^i : \mathcal{N} \to [0,1]$, where $\mathcal{N}(\mathbb{R})$ is the set of fuzzy numbers, gives for each i-th constraint the closeness of the two fuzzy numbers $\sum_{j=1}^{n} a_{ij} x_j$ and b_i in terms of a fuzzy number t_i (a margin of violation tolerance). In other words, the μ^i's play here the same role for the fuzzy coefficients as those in (29).

For solving (57), Delgago, Verdegay and Vila (1990) propose to replace the set of constraints by

$$\sum_{j=1}^{n} a_{ij} x_j > b_i + \underline{t}_i\,(1-a),\ \alpha \in [0,1],\ i=1, \ldots m \tag{58}$$

where > is some relation between two fuzzy numbers (just preserving the ranking under the multiplication by a positive scalar).

In this way the problem to be considered becomes

$$cx \to \max_{x \geq 0}$$

$$\text{subject to: } \sum_{j=1}^{n} a_{ij} x_j < b_i + t_i\,(1-\alpha),\ i = 1, \ldots, m \tag{59}$$

from which a fuzzy (optimal) solution of (57) can be obtained.

4. REMARKS ON DUALITY IN FUZZY LINEAR PROGRAMMING

Duality in fuzzy linear programming has been approached from various viewpoints. The first attempt should apparently be attributed to Rödder and Zimmermann (1980). Emphasis is there on a new economic interpretation of the classical minimax and maximin problems of the duality theory by using fuzzy sets for «softening» the problem's setting (e.g., the satisfaction of constraints).

Another approach by Hamacher, Leberling and Zimmermann (1978) starts from a (mixed) FLP problem involving both fuzzy and nonfuzzy constraints. Its solution is derived by solving an auxiliary LP problem of type (37) whose nonfuzzy dual is obtained, and then the dual variables are analyzed and interpreted.

Llena (1985) also presents an approach from the viewpoint of fuzzy multiobjective and fuzzy goal programming, and shows that one can define some «fuzzy duality» such that the equivalent forms obtained for the fuzzy primal and dual are also primal and dual in the classical sense.

A definition of the dual problem in the FLP, and a method for deriving it, given by Verdegay (1984), is based on a symmetry between the fuzzy constraints and fuzzy objective (cf. Section 2.1); the following two types of FLP problems are then considered:

a) Problem (P1) when the objective function is exactly known but the constraints are fuzzy; i.e. we obtain (29)
b) Problem (P2) when the constraints are exactly known but the (costs) coefficients in the objective function are fuzzy; i.e. we obtain (43).

Now, given an FLP problem of type (P1) and (P2), there always exists its dual, and they have both the same fuzzy solution. To be more specific, if we consider

$$\begin{aligned} & cx \to \max_{x} \\ & \text{subject to: } Ax < b,\ x \geq 0 \end{aligned} \tag{59}$$

and if μ_i, j = 1, 2, ..., m, the membership function of the j-th fuzzy constraint, is defined as

$$\mu_i(v) = \begin{cases} [(b_j + d_j) - v] / d_j & b_j + d_j \geq v \geq b_j \\ 1 & v \geq b_j \\ 0 & b_j + d_j < v \end{cases} \tag{61}$$

then it can be shown that the dual problem to (60) is

$$\begin{aligned} & \underset{\sim}{d}u \to \min_{u \geq 0} \\ & \text{subject to: } uA^T \geq c \end{aligned} \tag{62}$$

which is obviously an FLP problem with fuzzy coefficients d [given by (61)] in the objective function.

Both (60) and (62) have in each α-cut, $\alpha \in [0,1]$, the same fuzzy solution $z^*(\alpha) = c \cdot x^*(\alpha) = d(\alpha) \cdot u^*$ which is obtained (cf. Verdegay, 1984) using for (60) and (62) auxiliary parametric LP problems of type (40) and (46), respectively.

It is important to note (Verdegay, 1984) that for the FLP problems, with both fuzzy constraints and fuzzy coefficients in the objective function, the definition of their duals may be derived analogously as above.

5. REMARKS ON APPLICATIONS AND EXTENSIONS OF FUZZY LINEA PROGRAMMING

The FLP has found numerous applications in, e.g., designing the size and structure of a truck fleet (Zimmermann, 1976), media selection in advertising (Wiedey and Zimmermann, 1978) air pollution regulation (Sommer and Pollatschek, 1978), determination of agricultural policies (Kacprzyk and Owsinski, 1984; Owsinski, Zadrozny and Kacprzyk, 1987), etc.

Though the general FLP models may be applied in principe to any problem within their scope, some specific problems call for some special LP-type models, and their related special algorithms, more efficient that the general ones. A particular role is here played by the transportation problem fuzzified by, e.g., Verdegay (1984), Chanas, Kolodziejczyk and Machaj (1984) and Delgado, Verdegay and Vila (1987c). Moreover, fuzzy stochastic and possibilistic LP (e.g., Buckley, 1988, Lunhandjula, 1983, 1986), and fuzzy zero-sum games (e.g., Campos, 1989; Campos and Verdegay, 1986; Orlovski, 1977) may be cited.

For details on fuzzy integer and 0-1 programming, see Kacprzyk and Orlovski (1987b), and Verdegay and Herrera (in this volume).

6. CONCLUDING REMARKS

In this survey we have tried to present first some concepts, issues, tools, etc. which – in our opinion – are basic for broadly perceived fuzzy optimization and mathematical programming. Then, we concentrated on fuzzy linear programming due to its level of maturity and practical importance.

LITERATURE

BELLMAN R.E. and ZADEH L.A. (1970), *Decision Making in a Fuzzy Environment.* Man. Sci. 17 (B), 4, 141-164.

BUCKLEY J.J. (1983), *Fuzzy Programming and the Pareto Optial Set; Fuzzy Sets and Systems* 10, 57-63.

BUCKLEY J.J. (1988), *Possibility and Necessity in Optimization. Fuzzy Sets and Systems* 25, 1-13.

CAMPOS L. (1989), *Fuzzy Linear Programming Models to Solve Fuzzy Matrix Games.* Fuzzy Sets and Systems 32, 275-289.

CAMPOS L. and VERDEGAY J.L. (1986), *On fuzzy Games. Proc. of Fall International Seminar on Applied Logic.* Palma de Mallorca (Spain).

CAMPOS L. and VERDEGAY J.L. (1989), *Linear Programming Problems and Ranking of Fuzzy Numbers.* Fuzzy Sets and Systems 32, 1-11.

CAMPOS L., M. DELGADO and VILA M.A. (1990), *Solving Fuzzy Matrix Games by Using a Direct Approach.* In Verdegay and Delgado (1990), 69-84.

CARLSSON Ch. and KORHONEN P. (1986), *A parametric Approach to Fuzzy Linear Programming.* Fuzzy Sets and Systems 20, 17-30.

CHANAS S. (1983), *Parametric Programming in Fuzzy Linear Programming.* Fuzzy Sets and Systems, 11, 243-251.

CHANAS S. (1989), *Fuzzy Programming in Multiobjective Linear Programming – A Parametric Approach.* Fuzzy Sets and Systems 29, 303-313.

CHANAS S., KOLODZIEJCZYK W. and MACHAJ (1984), *A Fuzzy Approach to the Transportation Problem.* Fuzzy Sets and Systems 13, 211-222.

CHANAS S. and KULEJ M. (1984), *A Fuzzy Linear Programming Problem with equality constraints.* In Kacprzyk (1984). 195-202.

DELGADO M. (1983), *A Resolution Method for Multiobjective Problems.* Europ. J. of Op. Res. 13, 165-172.

DELGADO M., VERDEGAY J.L. and VILA M.A. (1985a), *Solving the Biobjective Linear Programming Problem: A Fuzzy Approach.* In Approximate Reasoning in Expert Systems (M.M. Gupta et al. Eds.), (1985) 317-322. North-Holland.

DELGADO M., VERDEGAY J.L. and VILA M.A. (1985b), *Fuzzy Vectormaximum Problem and Parametric Programming.* (In spanish). Trabajos de Estadistica y de Investigacion Operativa 36, 2, 126-137.

DELGADO M., VERDEGAY J.L. and VILA M.A. (1987a), *Imprecise Costs in Mathematical Programming Problems. Control and Cybernetics* 16, 113-121.

DELGADO M., VERDEGAY J.L. and VILA M.A. (1987b), *On Fuzzy Linear Programming Models. Preprints of II IFSA Congress.* Tokyo, 715-718.

DELGADO M., VERDEGAY J.L. and VILA M.A. (1987c), *Fuzzy Transporation Problems: A General Analysis.* In Kacprzyk and Orlovski (1987a), 342-358.

DELGADO M., VERDEGAY J.L. and VILA M.A. (1989a), *A General Model for Fuzzy Linear Programming.* Fuzzy Sets and Systems 29, 21-29.

DELGADO M., VERDEGAY J.L. and VILA M.A. (1989b), *Using Linguistic Labels in Games* (In Spanish). Proc. of the III Technical Meeting of the Spanish Association for Artificial Intelligence, 349-354.

DELGADO M., VERDEGAY J.L. and VILA M.A. (1990), *Relating Different Approaches to solve Linear Programming Problems with Imprecise Costs.* Fuzzy Sets and Systems 37, 33-42.

DUBOIS D. and PRADE H. (1980a): *Systems of Linear Fuzzy Constraints.* Fuzzy Sets and Systems 3, 37-48.

DUBOIS D. and PRADE H. (1980), *Fuzzy Sets and Systems. Theory and Applications.* Academic Press, New York.

DYSON R.G. (1980), *Maxmin Programming, Fuzzy Linear Programming and Multicriteria Decision Making.* J. of Oper. Res. Soc. 31, 263-267.

GARCIA-AGUADO M.C. (1990), *Sensitivity of Membership Functions in Fuzzy Linear Programming* (In Spanish). Ph. D. University of Granada.

HAMACHER H., LEBERLING H. and ZIMMERMANN H.J. (1978), *Sensitivity Analysis in Fuzzy Linear Programming.* Fuzzy Sets and Systems 1, 269-281.

HANNAN E.L. (1979), *On the Efficiency of the Product Operator in Fuzzy Programming with Multiple Objectives.* Fuzzy Sets and Systems 2, 259-262.

HANNAN E.L. (1981a), *Linear Programming with Multiple Fuzzy Goals.* Fuzzy Sets and Systems 6, 235-248.

HANNAN E.L. (1981b), *On Fuzzy Goal Programming.* Decision Sciences 12, 522-531.

HANNAN E.L. (1981b), *Contrasting Fuzzy Goal Programming and «Fuzzy» Multicriteria Programming.* Decision Sciences 13, 337-339.

IGNIZIO J.P. (1982), *On the (re) Discovery of Fuzzy Goal Programming Decision* Sciences 13, 331-336.

KABBARA G. (1982), *New Utilization of Fuzzy Optimization Method.* In M.M. Gupta and E. Sanchez (Eds), Fuzzy Information and Decision Processes, North-Holland, Amsterdam, 239-246.

KACPRZYK J. (1983), *Multistage Decision – Making under Fuzziness.* ISR Series, Verlag TÜV Rheinalnd, Cologne.

KACPRZYK J. (1984), *Guest Ed.: Special Issue on Fuzzy Sets and Possibility Theory in Optimization Models.* Control and Cybernetics 4, No. 3.

KACPRZYK J. and ORLOVSKI S.A., Eds (1987a), *Opimization Models Using Fuzzy Sets and Possibility Theory.* Reidel, Dordrecht.

KACPRZYK J. and ORLOVSKI S.A.(1987b), *Fuzzy Optimization and Mathematical Programming: A Brief Introduction and Survey*. In Kacprzyk and Orlovski (1987a).

KACPRZYK J. and YAGER R.R. (1984A), *«Softer» Optimization and Control Models via Fuzzy Linguistic Quantifiers*. Information Sciences 34, 157-178.

KACPRZYK J. and YAGER R.R. (1984b), *Linguistic Quantifiers and Belief Qualification in Fuzzy Multicriteria and Multistage Decision Making*. In Kacprzyk (1984), 155-174.

LAMATA M.T., MORAL S. and VERDEGAY J.L. (1990), *Transforming Fuzzy Measures*. In Verdegay and Delgado (1990), 146-158

LEBERLING H. (1981), *On finding Compromise Solutions in MultiCriteria Problems Using the Fuzzy Min-Operator*. Fuzzy Sets and Systems 6, 105-118.

LEUNG Y. (1982), *Multicriteria Conflict Resolution Through a Theory of Displaced Fuzzy Ideal*. In M.M. Gupta and E. Sanchez (Eds), Approximate Reasoning in Decision Analysis. North-Holland. 381-390.

LLENA J. (1985), *On Fuzzy Linear Programming*. Eur. J. Op. Res. 22, 216-223.

LUHANDJULA M.K. (1982), *Compensatory Operators in Fuzzy Linear Programming with Multiple Objectives*. Fuzzy Sets and Systems 8, 245-252.

LUHANDJULA M.K. (1983), *Linear Programming under Randomness and Fuzziness*. Fuzzy Sets and Systems 10, 57-63.

LUHANDJULA M.K. (1984), *Fuzzy Approaches for Multiple Objective Linear Fractional Optimization*. Fuzzy Sets and Systems 13, 11-23.

LUHANDJULA M.K. (1986), *Satisfying Solutions for a Possibilistic Linear Program*. Information Sciences 40, 247-265.

LUHANDJULA M.K. (1987a), *Multiple Objective Programming Problems with Possibilistic Coefficients*. Fuzzy Sets and Systems 21, 135-146.

LUHANDJULA M.K. (1987), *Linear Programming with a Possibilistic Objective Function*. Eur. J. of Op. Res. 31, 110-117.

LUHANDJULA M.K. (1989), *Fuzzy Optimization: an appraisal*. Fuzzy Sets and Systems 30, 257-282.

NAKAMURA K. (1984), *Some Extension of Fuzzy Linear Programming*. Fuzzy Sets and Systems 14, 211-229.

NARASIMHAN R. (1980), *Goal Programming in a Fuzzy Environment*. Decision Sci. 11, 325-336.

NARASIMHAN R. (1981), *On Fuzzy Goal Programming - Some Comments*. Decision Sci. 12, 532-538.

NEGOITA C.V. (1981), *The Current Interest in Fuzzy Optimization*. Fuzzy Sets and Systems 6, 261-269.

NEGOITA C.V. (1984), *Structure and Logic in Optimization*. In Kacprzyk (1984), 121-128.

NEGOITA C.V., FLONDOR P. and SULARIA M. (1977), *On Fuzzy Environment in Optimization Problems*. In Modern Trends in Cybernetic and Systems. J. Rose and C. Bilciu (Eds). Springer-Verlag.

NEGOITA C.V., S. MINOIU and STAN E. (1976), *In Considering Imprecision in Dynamic Linear Programming*. Economic Computation and Economic Cybernetics Studies and Research 3, 83-95.

NEGOITA C.V. and RALESCU D. (1975), *Applications of Fuzzy Sets to Systems Analysis*. Birkhauser-Verlag.

NEGOITA C.V. and RALESCU D. (1977), *On Fuzzy Optimization*. Kybernetes, 6, 193-195.

NEGOITA C.V. and STEFANESCU A.C. (1982), *On Fuzzy Optimization*. In M.M. Gupta and E. Sanchez (Eds), Fuzzy Information and Decision Processes. North-Holland, Amsterdam, 247-250.

NEGOITA C.V. and SULARIA M. (1976): *On Fuzzy Programming and Tolerances in Planning*. Econom. Comp. Econom. Cybernet. Stud. Res., 1, 3-15.

OH'EIGEARTAIGH M. (1982), *A Fuzzy Transportation Algorithm*. Fuzzy Sets and Systems, 8, 235-245.

ORLOVSKI S.A. (1977), *On Programming with Fuzzy Constraint Sets*. Kybernetes 6, 197-201.

ORLOVSKI S.A. (1978), *Decision Making with a Fuzzy Preference Relation*. Fuzzy Sets and Systems 1, 155-167.

ORLOVSKI S.A. (1980), *On Formulization of a General Fuzzy Mathematical Problem*. Fuzzy Sets and Systems 3, 311-321.

ORLOVSKI S.A. (1982), *Effective Alternatives for Multiple Fuzzy Preference Relations*. In R. Trappl (Ed.), Cybvernetics and Systems Research. North-Holland, Amsterdam.

ORLOVSKI S.A. (1984), *Multiobjective Programming Problems with Fuzzy Parameters*. In Kacprzyk (1984), 175-184.

RALESCU D. (1984), *Optimization in a Fuzzy Environment*. In M.M. Gupta and E. Sanchez (Eds), Fuzzy Information, Knowledge Representation and Decision ANalysis. Pergamon Press, Oxford.

RAMIK J. (1983), *Extension Principle and Fuzzy Mathematical Programming*. Kybernetika 19, 516-525.

RAMIK J. (1986), *Extension Principle in Fuzzy Optimization*. Fuzzy Sets and Systems 19, 29-36.

RÖDDER W. and H.J. ZIMMERMANN (1980): *Duality in Fuzzy Linear Programming*. In A.V. Fiacco and K.O. Kortanek (Eds), Extremal Methods and Systems Analyses. Springer-Verlag. Berlin, 415-429.

ROMMELFANGER, H., R. HANUSCHECK and WOLF J. (1985), *Linear Programming with Fuzzy objectives*. Fuzzy Sets and Systems 29, 31-48.

RUBIN P.A. and NARASIMHAN R. (1984), *Fuzzy Goal Programming with Nested Priorities*. Fuzzy Sets and Systems 14, 115-129.

SLOWINSKI R. (1986), *A Multicriteria Fuzzy Linear Programming Method for Water Supply System Development Planning*. Fuzzy Sets and Systems 19, 217-238.

SAKAWA M. (1983), *Interactive Computer Programs for Fuzzy Linear Programming with Multiple Objectives*. Int. J. of Man-Machine Studies 18, 489-503.

SAKAWA M. (1984a), *Interactive Fuzzy Goal Programming for Multiobjective Nonlinear Problems and its Application to Water Quality Management*. In Kacprzyk (1984), 217-228.

SAKAWA M. (1984b), *Interactive Multiobjective Decision Making by the Fuzzy Sequential Proxy Optimization Technique: FSPOT*. In Zimmermann, Zadeh and Gaines (1984), 241-260.

SAKAWA M. and F. SEO (1983), *Interactive Multiobjective Decision Making in Environmental Systems using the Fuzzy Sequential Proxy Optimization Technique*. Large Scale Systems 4, 223-243.

SAKAWA M. and YANO H. (1985), *Interactive Decision Making for Multiobjective Linear Fractional Programming Problems with Fuzzy Parameters*. Cybernetics and Systems 16, 377-394.

SAKAWA M., YANO H. and YUMINE T. (1987), *An Interactive Fuzzy Satisficing Method for Multiobjective Linear Programming Problems and its Application.* IEEE Transactions on Systems, Man and Cybernetics SMC - 17, 654-651.

SAKAWA M., YUMINE T. (1983), *Interactive Fuzzy Decision Making for Multiobjective Linear Fractional Programming Problems.* Large Scale Systems 5, 105-114.

SHER A.P. (1980), *Solving Mathematical Programming Problem with a Linear Goal Function in Fuzzy Constraints (in Russian).* Automation and Remote Control 40, 137-143.

SOMMER G. and POLLATSCHEK M.A. (1978), *A Fuzzy Programming Approach to an Air Pollution Regulation Problem.* Eur. J. Op. Res. 10, 303-313.

TAKEDA E. and NISHIDA N.T. (1980), *Multiple Criteria Decision Making with Fuzzy Domination Structures.* Fuzzy Sets and Systems 3, 123-136.

TANAKA H. and ASAI K, (1981), *Fuzzy Linear Programming Based on Fuzzy Functions.* Proc. 8th IFAC World Congress (Kyoto, Japan), Pergamon Press, Oxford.

TANAKA H. and ASAI K. (1984a), *Fuzzy Linear Programming Problems with Fuzzy Numbers.* Fuzzy Sets and Systems 13, 1-10.

TANAKA H. and ASAI K. (1984b): *Fuzzy Solution in Fuzzy Linear Programming Problems.* IEEE Trans. on Systems, Man and Cybernetics SMC-14, 285-288.

TANAKA H., ICHIHASHI H. and ASAI K. (1984), *A Formulation of Fuzzy Linear Programming Problems based on Comparison of Fuzzy Numbers.* In Kacprzyk (1984), 185-194.

TANAKA H., OKUDA T. and ASAI K. (1974), *On Fuzzy Mathematical Programming.* Journal of Cybernetics, 3, 37-46.

VERDEGAY J.L. (1982): *Fuzzy Mathematical Programming.* In Fuzzy Information and Decision Processes (M.M. Gupta and E. Sanchez, Eds.). North-Holland, 231-237.

VERDEGAY J.L. (1983), *Transportation Problems with Fuzzy Parameter.* (In Spanish). Rev. Acad. Ciencias Mat. Fis. Quim y Nat. De Granada, 2, 47-56.

VERDEGAY J.L. (1984a), *A Dual Approach to Solve the Fuzzy Linear Programming Problem.* Fuzzy Sets and Systems 14, 131-141.

VERDEGAY J.L. (1984b), *Application of Fuzzy Optimization in Operational Research.* In Kacprzyk (1984) 229-239.

VERDEGAY J.L. (1987), *Fuzzy Mathematical Programming Problem: Resolution.* In M.G. Singh (Ed.), Systems and Control Encyclopedia. Theory, Technology, Applications. Pergamon Press, 1816-1819.

VERDEGAY J.L. and DELGADO M. (1989), *The Interface between Artificial Intelligence and Operations Research in Fuzzy Environment.* Verlag TÜV Rheinland, ISR Series no. 95.

VERDEGAY J.L. and DELGADO M. (1990), *Approximate Reasoning Tools for Artificial Intelligence.* Verlag TÜV Theinland, ISR Series no. 96

WERNERS B. (1987), *Interactive Multiple Objective Programming Subject to Flexible Coefficients.* Europ. J. of Op. Res. 31, 342-349.

WIEDEY G. and ZIMMERMANN H.J. (1978), *Media Selection and Fuzzy Linear Programming.* J. Op. Res. Soc. 29, 1071-1084.

WIERZCHON S.T. (1987), *Linear Programming with Fuzzy Sets: A General Approach.* Mathematical Modelling 9, 447-460.

YAGER R.R. (1979), *Mathematical Programming with Fuzzy Constraints and a Preference on the Objective.* Kybernetes 9, 109-114.

ZIMMERMANN H.J. (1975), *Description and Optimization of Fuzzy Systems.* Int. J. of General Systems 2, 209-215.

ZIMMERMANN H.J. (1978), *Fuzzy Programming and Linear Programming with Several Objective Functions.* Fuzzy Sets and Systems 1, 45-55.

ZIMMERMANN H.J. (1983), *Using Fuzzy Sets in Operational Research.* European Journal of Operational Research 13, 201-216.

ZIMMERMANN H.J. (1985), *Applications of Fuzzy Sets Theory to Mathematical Programming.* Information Sciences 36, 29-58.

2. FUZZY OPTIMIZATION: GENERAL ISSUES AND RELATED TOPICS

MINIMIZING A FUZZY FUNCTION

Elio Canestrelli and Silvio Giove
Department of Applied Mathematics and Computer Science
University of Venice
Dorsoduro 3825/E
Venice, Italy

Abstract: The present paper considers the unconstrained optimization of a continuous fuzzy function with respect to fuzzy numbers of the L-R type. We extend some methods of solution of non-fuzzy parametric mathematical programming to the fuzzy case.

Keywords: fuzzy number, fuzzy optimization.

1. INTRODUCTION

In a previous paper (Canestrelli and Giove, 1989) we faced the problem of the minimization of a fuzzy quadratic function:

$$f(\tilde{\mathbf{x}},\tilde{\mathbf{b}}) = \frac{1}{2}\,\tilde{\mathbf{x}}'A\,\tilde{\mathbf{x}} + \tilde{\mathbf{b}}'\tilde{\mathbf{x}}\ ,$$

with A being positive definite, showing that, similarly to the deterministic case, the minimizing vector is $\tilde{\mathbf{x}}_0 = -A^{-1}\tilde{\mathbf{b}}$ and the minimum is $\tilde{y}_0 = -\frac{1}{2}\,\tilde{\mathbf{b}}'A^{-1}\tilde{\mathbf{b}}$. In the present paper we demonstrate that the previously obtained result can easily be extended to any kind of continuous function.

Let be $f(\mathbf{x},\mathbf{p})$ a nonfuzzy function with $\mathbf{x}\in\mathbb{R}^n$, $\mathbf{p}\in\mathbb{R}^n$ and $f:\mathbb{R}^{n+m}\to\mathbb{R}$. Let us suppose that, for any vector $\mathbf{p}\in\mathbb{R}^m$, the vector

$$\mathbf{x}_0 = g(\mathbf{p}) \qquad (1)$$

is the unique solution of the problem

$$z_0 = f(\mathbf{x}_0,\mathbf{p}) \qquad (2)$$

with $g:\mathbb{R}^m\to\mathbb{R}^n$. Let be $f(\tilde{\mathbf{x}},\tilde{\mathbf{p}})$ the fuzzy extension of the nonfuzzy function $f(\mathbf{x},\mathbf{p})$ (Dubois and Prade, 1980, p.98). Besides, let us suppose the fuzzy numbers p_i $(i=1,2,\dots,m)$ to be noninteractive (Dubois and Prade, 1980, p.70).

In the present paper we intend to demonstrate that the

minimizing vector $\tilde{\mathbf{x}}_0$, which solves the fuzzy problem (2), is obtained in a way formally similar to the nonfuzzy case by calculating $\tilde{\mathbf{x}}_0=g(\tilde{\mathbf{p}})$ and successively, if requested, $\tilde{z}_0=f(\tilde{\mathbf{x}}_0,\tilde{\mathbf{p}})$.

In what follows we shall suppose the following hypotheses as verified:

a. the components of vector $\tilde{\mathbf{p}}$ are L-R fuzzy numbers whose support is a closed real interval and whose membership function is continuous. We shall represent a generic component of $\tilde{\mathbf{p}}$ by a triple (c,a_l,a_r);
b. the function $f(\mathbf{x},\mathbf{p})$ is everywhere continuous with respect to the (n+m) dimensional vector $(\mathbf{x},\mathbf{p})$.

2. FUZZY MINIMIZATION

Let us suppose the interactivity relation between fuzzy numbers to be known, with a special reference to a functional link of type $\tilde{\mathbf{x}}=f(\tilde{\mathbf{p}})$.

Given two real intervals I_1 and I_2, the partial order relation $I_1 \ll I_2$ holds if and only if the two following relations hold:

$$\inf I_1(h) \le \inf I_2(h)$$
$$\sup I_1(h) \le \sup I_2(h).$$

Given two fuzzy numbers $\tilde{x}$ and $\tilde{y}$, we say that $\tilde{x}\le\tilde{y}$ if and only if $\tilde{x}=\min(\tilde{x},\tilde{y})$ (Dubois and Prade, 1980, p. 52 and Tanaka, Ichihashi, Asai, 1984, p. 188). Such a partial order relation leads to the following relation between h-cut intervals

$$I_x(h) \ll I_y(h), \qquad 0\le h\le 1 \tag{3}$$

where

$$I_x(h)=\{a\in \mathrm{Supp}(\tilde{x}):\mu_x(a)\ge h\}. \tag{4}$$

So, we note that, if in the nonfuzzy case for any choice of vector $\mathbf{p}$ the minimizing vector of (2) is expressed by (1), then for every function $v:\mathbb{R}^m\to\mathbb{R}^n$ it results

$$f(g(\mathbf{p}),\mathbf{p})\le f(v(\mathbf{p}),\mathbf{p}). \tag{5}$$

Or, if we let

$$\alpha(\mathbf{p})\equiv f(g(\mathbf{p}),\mathbf{p}) \qquad \text{and} \qquad \beta(\mathbf{p})\equiv f(v(\mathbf{p}),\mathbf{p}) \tag{6}$$

we obtain

$$\alpha(\mathbf{p})\le\beta(\mathbf{p}) \qquad \text{for any } \mathbf{p}\in\mathbb{R}^m. \tag{7}$$

Proceeding to the fuzzy case, let us introduce the fuzzy function $f(\tilde{\mathbf{x}},\tilde{\mathbf{p}})$, assume the validity of the hypotheses a and b of section 1, and give some explanations.

In a previous work, we examined some properties of

interactive and linked fuzzy numbers (Canestrelli and Giove, 1989). In particular, we noted that, if $\tilde{u}=\gamma(\tilde{v})$, the relation between the fuzzy numbers $\tilde{u}$ and $\tilde{v}$ is a more restrictive concept than the fuzzy interactivity, so we introduced the expression: ***$\tilde{u}$ and $\tilde{v}$ linked by γ***. Such a relation between $\tilde{u}$ and $\tilde{v}$ implies that, for any $v\in Supp(\tilde{v})$, we have to assign to the membership function of the couple $(\gamma(\tilde{v}),\tilde{v})$ the value $\mu_v(v)$, i.e. the same value of the membership function of $\tilde{v}$. This is coherent with an intuitive meaning of a fuzzy number. As remarked in the former reference, if v is the value of the size of an object, whatever its value may be, it is obvious, for example, that the difference v-v is always zero (crisp).

The same result must be applied to any function $g(\tilde{\mathbf{p}})$, so that, for every function g, we have to assign to $\tilde{y}=f(g(\tilde{\mathbf{p}}),\tilde{\mathbf{p}})$ the following membership function

$$\mu_y(a)=\sup_{\mathbf{p}\in\mathbb{R}^m}\{\min[\mu_{p_1}(p_1),\ldots,\mu_{p_m}(p_m)]:a=f(g(\mathbf{p}),\mathbf{p})\}. \qquad (8)$$

For what is said above, the minimizing value $\tilde{\mathbf{x}}_0$ of a parametric function $f(\tilde{\mathbf{x}},\tilde{\mathbf{p}})$ must be a fuzzy function of the fuzzy parameter $\tilde{\mathbf{p}}$. In the following, we will see that this function is $\tilde{\mathbf{x}}_0=g(\tilde{\mathbf{p}})$.

Let us note, first, that the function $\alpha(\mathbf{p})$, the minimum with respect to $\mathbf{x}$ of $f(\mathbf{x},\mathbf{p})$, is a continuous function in $\mathbf{p}$, f being continuous in $(\mathbf{x},\mathbf{p})$. Then, let us set

$$\alpha(\tilde{\mathbf{p}})=f(g(\tilde{\mathbf{p}}),\tilde{\mathbf{p}}) \qquad \text{and} \qquad \beta(\tilde{\mathbf{p}})=f(v(\tilde{\mathbf{p}}),\tilde{\mathbf{p}}) \qquad (9)$$

and verify that $\alpha(\tilde{\mathbf{p}})$ is the minimum of $f(\tilde{\mathbf{x}},\tilde{\mathbf{p}})$. To this purpose, it is sufficient to show that (3), with $\tilde{\mathbf{x}}=\alpha(\tilde{\mathbf{p}})$ and $\mathbf{y}=\beta(\mathbf{p})$, is valid. In other words, if we set:

$$U_\alpha(h)=\{b\in\mathbb{R}\mid \alpha(\mathbf{p})=b,\ \mathbf{p}\in I_p(h)\} \qquad \text{and}$$

$$U_\beta(h)=\{b\in\mathbb{R}\mid \beta(\mathbf{p})=b,\ \mathbf{p}\in I_p(h)\},$$

it has to imply

$$U_\alpha(h) \ll U_\beta(h). \qquad (10)$$

The function $\alpha(\mathbf{p})$ maps the closed bounded m-dimensional interval $I_p(h) = I_{p_1}(h) \times I_{p_2}(h) \times\ldots\times I_{p_m}(h)$ in the closed and bounded interval $U_\alpha(h)$.

If we get $p_\alpha^h \in I_p(h)$ and $p_\beta^h \in I_p(h)$ in such way that

$$\alpha(p_\alpha^h) = \inf U_\alpha(h) \qquad \text{and} \qquad \beta(p_\beta^h) = \inf U_\beta(h),$$

then from (7) we obtain $\alpha(p_\beta^h) \le \beta(p_\beta^h)$.

Consequently

$$\inf U_\alpha(h) \le \inf U_\beta(h). \qquad (11)$$

Similarly, we verify that

$$\sup U_\alpha(h) \le \sup U_\beta(h). \tag{12}$$

As (11) and (12) have to be valid for every value of h, $0 \le h \le 1$, that is sufficient to affirm, through (3) and (4), that $\tilde{\mathbf{x}}_0 = g(\tilde{\mathbf{p}})$ is a minimizing vector of the fuzzy function $f(\tilde{\mathbf{x}}, \tilde{\mathbf{p}})$, and it membership function is

$$\mu_{x_0}(a) = \sup_{\mathbf{p} \in \mathbb{R}^m} \{\min[\mu_{p_1}(p_1), \ldots, \mu_{p_m}(p_m)] : a = g_1(\mathbf{p})\}. \tag{13}$$

In general, despite of the continuity of the membership function of the components of $\tilde{\mathbf{p}}$ and of $f(\tilde{\mathbf{x}}, \tilde{\mathbf{p}})$, neither the continuity of the membership function of $\tilde{\mathbf{x}}_0$, nor of $\tilde{y}_0 = f(\tilde{\mathbf{x}}_0, \tilde{\mathbf{p}})$ is guaranteed.

Besides, special care is needed to calculate $\tilde{\mathbf{x}}_0$ and $\tilde{y}_0$ because we are dealing with operations with interactive numbers. Particularly, we note that when computing $\tilde{y}_0 = f(\tilde{\mathbf{x}}_0, \tilde{\mathbf{p}})$ we must consider a relation between $\tilde{\mathbf{x}}_0$ and $\tilde{\mathbf{p}}$, $\tilde{\mathbf{x}}_0 = g(\tilde{\mathbf{p}})$, which gives a "link" between the components of vector $\tilde{\mathbf{x}}_0$. In a previous paper (Canestrelli and Giove, 1989) we remarked that in the presence of interactive fuzzy numbers, it is not allowed to use the calculation rules developed for increasing binary operations (Dubois and Prade, 1980 p. 42). Their application, in fact, is valid if the operands are not interactive.

3. AN EXAMPLE

Let us consider the nonfuzzy function $f: \mathbb{R} \to \mathbb{R}$:

$$y = f(x,p) = \begin{cases} e^{\pi p x - \sin(\pi p)} - \pi p x & \text{if} \quad p \neq 0 \\ (x-1)^2 & \text{if} \quad p = 0. \end{cases} \tag{14}$$

The unique minimizing value x_0 is obtained from the equation $y'_x = 0$, i. e. from:

$$\begin{aligned} & e^{\pi p x - \sin(\pi p)} = 1 && \text{if} \quad p \neq 0 \\ & x = 1 && \text{if} \quad p = 0. \end{aligned}$$

Solving it with respect to x, we obtain

$$x_0 = \mathrm{sinc}(p) \qquad \text{and} \qquad y_0 = 1 - \sin(\pi p),$$

where the function sinc(z) is defined in the following way:

$$\mathrm{sinc}(z) = \begin{cases} \dfrac{\sin(\pi z)}{\pi z} & \text{if} \quad z \neq 0 \\ 1 & \text{if} \quad z = 0. \end{cases}$$

Using the preceding results, it follows that the minimizing value and the minimum of the fuzzy extension of the nonfuzzy function

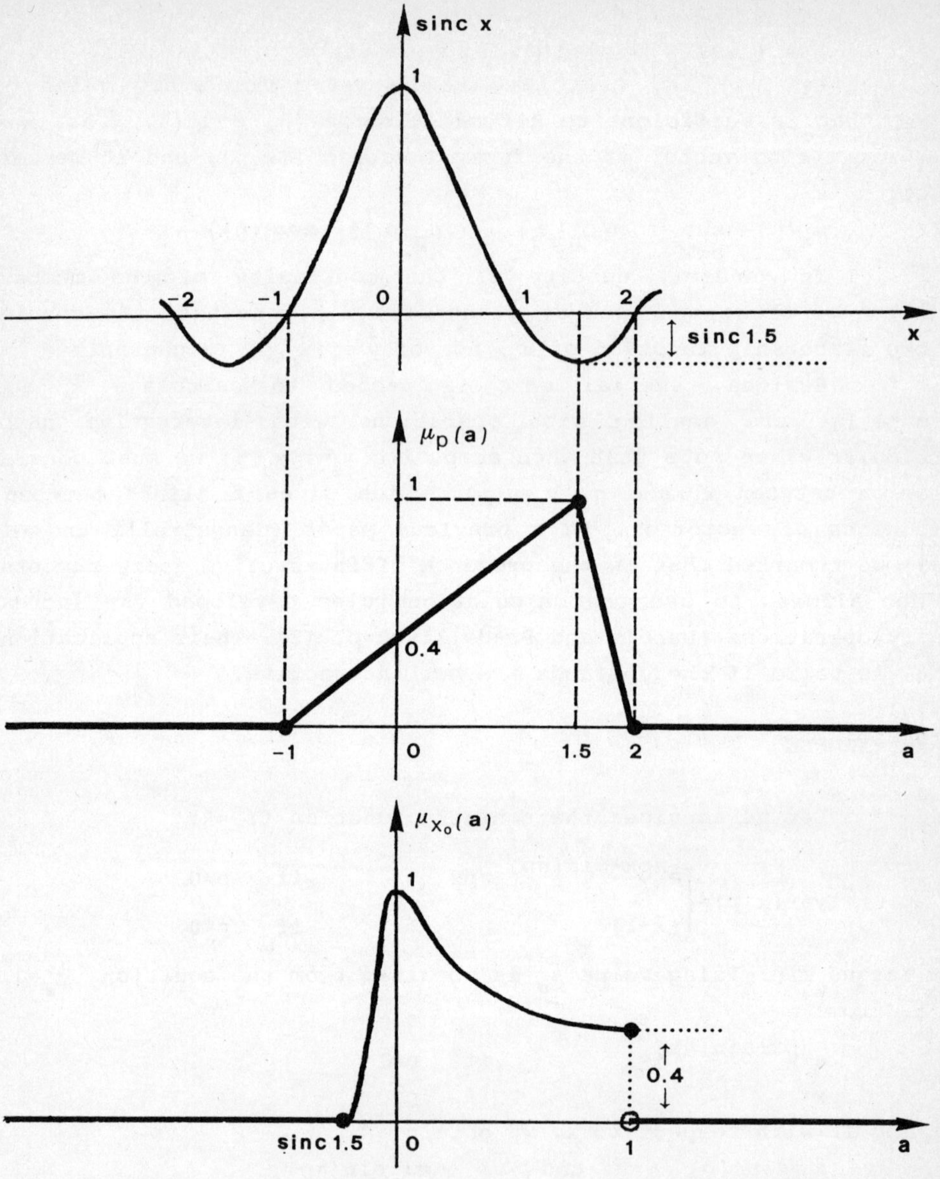

Fig. 1. An example

(14) are $\tilde{x}_0$=sinc($\tilde{p}$) and $\tilde{y}_0$= 1-sin($\pi\tilde{p}$).

The calculation of the membership functions of $\tilde{x}_0$ and $\tilde{y}_0$ is rather complicated. In fact, we must calculate respectively:

$$\mu_{x_0}(a) = \sup_{p\in\mathbb{R}} \{\mu_p(p) \mid a=\mathrm{sinc}(p)\},$$

$$\mu_{y_0}(a) = \sup_{p\in\mathbb{R}} \{\mu_p(p) \mid a=1-\sin(p)\}.$$

In both cases we can apply the algorithm we proposed in a previous paper (Canestrelli and Giove, 1990).

In Fig. 1 there is an approximate graph of the membership function of $\tilde{x}_0$, given the triangular fuzzy number $\tilde{p}$=(1.5, 2.5, 0.5). We can notice the existence of a first kind discontinuity in correspondence with the value $\bar{y}$=sinc(1.5), due the non-monotonicity of the function sinc(p).

REFERENCES

Canestrelli E., and S. Giove (1989) Optimizing a quadratic Function with fuzzy linear coefficients. in Proceedings of XIII AMASES Meeting, Verona (Italy) (to appear also in Control and Cybernetics, 1-1992)

Canestrelli E., and S. Giove (1990) An Algorithmic Approach to fuzzy Functions determination. to appear in Proceedings of the 8th Italian-Polish Symposium, Levico T.(TN), Italy.

Dubois D., and H. Prade (1980) Fuzzy Sets and Systems: Theory and Applications. Academic Press, New York.

Tanaka H., Ichihashi H. and K. Asai (1984) A formulation of fuzzy linear programming problem based on comparison of fuzzy numbers. Control and Cybernetics 13, 185-194.

A CONCEPT OF OPTIMALITY FOR FUZZIFIED MATHEMATICAL PROGRAMMING PROBLEMS*

Margit Kovács
Computer Center, Eötvös Loránd University
P.O.Box 157, H-1502,Budapest 112
Hungary

Abstract: In this paper an optimality concept is introduced for *(g,p)* - fuzzified mathematical programming problems. It is based on the parametric embedding method of classical nonlinear programming. A numerical method to find the optimum in this sense is also given.

Keywords: fuzzification, mathematical programming, parametrical embedding method, penalty function method.

1. INTRODUCTION

Let us consider the classical mathematical programming problem

$$f_0(\alpha_0, x) \longrightarrow \min \tag{1}$$

subject to

$$C = \{x \in I\!R^n_+ : f_i(\alpha_i, x) \le \alpha_{i0},\ i = 1, \ldots, m,\ f_i(\alpha_i, x) = \alpha_{i0},\ i = m+1, \ldots, s\}, \tag{2}$$

where $\alpha_i \in I\!R^n$, $i = 0, 1, \ldots, s$, $\alpha_{i0} \in I\!R$, $i = 1, \ldots, s$, and the parametrical functions $f_i(\alpha_i, x)$ are convex in x and linearly depend on the parameters, i.e. $f_i(\alpha_i, x) = \sum_{j=1}^{r} \alpha_{ij} h_j(x)$, $i = 0, 1, \ldots, s$.

In problem (1)–(2) the vector-parameters $\alpha_i \in I\!R^r$ and the right-hand-side parameters α_{i0} are supposed to be well defined characteristics of the problem modelled. However, these parameters are generally known only approximatively.

In this paper instead of (1)–(2) we will examine the fuzzified version of (1)–(2) assuming that every coordinate of the parameters in the problem formulation is given by fuzzy numbers. The support set of the fuzzy numbers describes the set of possible values of the perturbed parameters. The fuzzification concept is based on the idea that the uncertainty in the real world has a functional causality and the same functional causality takes effect in the choice of parameters as well as in the arithmetic operations. The optimum concept is motivated by the parametrical embedding method for the classical nonlinear programming (Vasil'ev, 1978).

* This research was supported by the projects OTKA-606-1986/400-0413 and FKP-400-0113

2. PRELIMINARIES

Let X be a universe and I be the unit interval of the real line $\mathbb{R}$. Denote $\mathcal{F}(X) = \{\mu| : X \to I\}$ the set of all *fuzzy subsets* on X. The characteristic function of $A \subset X$ as a special fuzzy subset will be denoted by χ_A.

A binary operation $T : I^2 \to I$ is said to be a *t-norm* iff it is commutative, associative, non-decreasing and $T(a,1) = a$ for all $0 < a < 1$. A t-norm is Archimedean iff it admits the representation

$$T(a,b) = g^{(-1)}\big(g(a) + g(b)\big)$$

where the *generator function* $g : I \to \mathbb{R}_+$ is continuous, strictly decreasing, $g(1) = 0$ and $g^{(-1)}(x) = g^{-1}(x)$ if $x \in [0, g(0)]$ and $g^{(-1)}(x) = 0$ for all $x \geq g(0) = g_0$ $\quad$ ($g_0 = \infty$ is also allowed). Every t-norm induces an n–ary operation $T^{n-1} : I^n \to I$ with the following rule

$$T^{n-1}(a_1, a_2, \ldots, a_n) = T\big(T^{n-2}(a_1, a_2, \ldots, a_{n-1}), a_n\big).$$

If T is Archimedean then

$$T^{n-1}(a_1, a_2, \ldots, a_n) = g^{(-1)}\Big(\sum_{j=1}^{n} g(a_j)\Big).$$

The *T-intersection* and *T-Cartesian product* of fuzzy sets will be defined as follows: if $\mu, \nu \in \mathcal{F}(X)$ then $(\mu \wedge_T \nu)(x) = T(\mu(x), \nu(x))$; if $\mu \in \mathcal{F}(X), \nu \in \mathcal{F}(Y)$ then $(\mu \times_T \nu)(x,y) = T(\mu(x), \nu(y))$.

Let $g : I \to \mathbb{R}_+$ be a fixed function with the properties of a generator function and let $\mathcal{F}_g$ denote the subset of $\mathcal{F}(\mathbb{R})$ containing the fuzzy numbers with the membership function

$$\mu(a) = \begin{cases} g^{(-1)}\big(|a - \alpha|/d\big) & \text{if } d > 0 \\ \chi_{\{\alpha\}}(a) & \text{if } d = 0 \end{cases}$$

for all $\alpha \in \mathbb{R}, d \in \mathbb{R}_+ \cup \{0\}$. The elements of $\mathcal{F}_g$ will be called *quasitriangular fuzzy numbers* generated by g with the center α and width d and we will recall for it by the pair (α, d).

Let T_{gp} be an Archimedean t-norm given by the generator function $g^p, 1 \leq p < \infty$. It is easy to see that $\lim_{p \to \infty} T_{gp}(a,b) = \min(a,b)$, therefore we will also use the notation T_{gp} in the case $p = \infty$ meaning the min-norm for $T_{g\infty}$.

The T_{gp}–Cartesian product of r quasitriangular fuzzy numbers generated by g will be called *(g,p)–fuzzy vector* on $\mathcal{F}_g^r$, i.e. $\mu_a = \mu_1 \times \ldots \times \mu_r$, where $\mu_i = (\alpha_i, d_i) \in \mathcal{F}_g$, $i = 1, \ldots, r$. Shortly we will also use the notion $\mu_a = (\alpha, d) \in \mathcal{F}_g^r$, meaning $\alpha \in \mathbb{R}^r, d \in \mathbb{R}^r$. It is obvious that $\mu_a \in \mathcal{F}_g^r \subset \mathcal{F}(\mathbb{R}^r)$. Without losing the generality we can assume that $d_i \neq 0$ for $i = 1, \ldots, k$ and $d_i = 0$ for $i = k+1, \ldots, r$, where the cases $k = 0$ and $k = r$ are also allowed. It is easy to show that

$$\mu_a(a) = \mu_a(a_1, \ldots, a_r) = \begin{cases} g^{(-1)}\big(\|\hat{D}^{-1}(\hat{a} - \hat{\alpha})\|_p\big), & \text{if } k = r \text{ or } a_i = \alpha_i,\ i = k+1, \ldots, r, \\ 0, & \text{otherwise} \end{cases} \qquad (3)$$

if $k > 0$, where (and everywhere in the sequel) $\hat{D} = \text{diag}(d_1, \ldots, d_k)$, $\hat{a} = (a_1, \ldots, a_k)$, $\hat{\alpha} = (\alpha_1, \ldots, \alpha_k)$ and

$$\|a\|_p = \begin{cases} \left(\sum_{j=1}^{k} |a_j|^p\right)^{1/p}, & \text{if } 1 \le p < \infty \\ \max_{j=1,\ldots,k} |a_j|, & \text{if } p = \infty \end{cases}$$

If $k = 0$, then it is obvious that

$$\mu_a(a) = \mu_a(a_1, \ldots, a_r) = \chi_{\{\alpha_1,\ldots,\alpha_r\}}(a_1, \ldots, a_r). \tag{4}$$

3. FUZZIFIED FUNCTIONS AND RELATIONS

3.1. Fuzzification of function values

At every fixed $x \in \mathbb{R}^n$ a *(g,p)-fuzzification* of the parametrical function value $f(a, x) = f(a_1, a_2, \ldots, a_r, x)$ by the *(g,p)*-fuzzy parameter vector $\mu_a = (\alpha, d) \in \mathcal{F}_g^r$ is a fuzzy set on $\mathbb{R}$:

$$\tilde{f}_T(\mu_a, x)(y) = \begin{cases} \sup_{a \in A(x,y)} \mu_a(a), & \text{if } A(x,y) \neq \emptyset, \\ 0, & \text{if } A(x,y) = \emptyset, \end{cases} \tag{5}$$

where $A(x, y) = \{a = (a_1, \ldots, a_r) \,|\, a_i \in \mathbb{R},\ y = f(a, x)\}$.

In the papers Kovács (1989) and Kovács (1991) has been proved that the *(g,p)*-fuzzified function values of a linear function $\ell(a, x)$ are quasitriangular fuzzy numbers $(\ell(\alpha, x), d\|x\|_q)$ assuming that every coordinate of the fuzzifying *(g,p)*-fuzzy vector has the same width $d > 0$, and $q = \begin{cases} 1, & \text{if } p = \infty \\ \infty, & \text{if } p = 1 \\ p/(p-1), & \text{otherwise} \end{cases}$. In Kovács (1990) this result has been generalized pointing to that the fuzzified function value will be modified only in its widths if the coordinates of the fuzzifying vector have different widths, namely, in this case the *(g,p)*-fuzzified function values are $(\ell(\alpha, x), \|Dx\|_q)$, where D is a diagonal matrix made of the widths of the fuzzifying vector. In both cases it was assumed that every parameter has positive perturbation possibility for the fuzzification, i.e. $d_i > 0$ for every i.

Now let us consider the nonlinear cases with arbitrary fuzzifying *(g,p)*-vectors, when the parametrical function linearly depends on the parameters.

Theorem 3.1.1. *Let the function $f(a, x) = \sum_{j=1}^{r} a_j h_j(x)$ be fuzzified by the (g,p)-fuzzy vector parameter $\mu_a = (\alpha, d) \in \mathcal{F}_g^r$. If $d_j > 0$, $j = 1, \ldots, k$, $d_j = 0$, $j = k+1, \ldots, r$, then the (g,p)-fuzzified function value is the fuzzy number $\tilde{f}(\mu_a, x) = (f(\alpha, x), \|Dh(x)\|_q)$, i.e. if $k > 0$ then*

$$\tilde{f}(\mu_a, x)(y) = \begin{cases} g^{(-1)}\left(|y - f(\alpha, x)| / \|\hat{D}\hat{h}(x)\|_q\right), & \text{if } \hat{h}(x) \neq 0 \\ \chi_{\{0\}}(y), & \text{if } \hat{h}(x) = 0 \end{cases} \tag{6}$$

and if $k = 0$ then

$$\tilde{f}(\mu_a, x)(y) = \chi_{\{f(\alpha,x)\}}, \tag{7}$$

where $D = \operatorname{diag}(d_1, \ldots, d_r)$, $\hat{D} = \operatorname{diag}(d_1, \ldots, d_k)$, $h(x) = (h_1(x), \ldots, h_r(x))$, $\hat{h}(x) = (h_1(x), \ldots, h_k(x))$.

Proof. The case $k = 0$ is trivial from (4) and (5). Assume now that $k > 0$. Then using (3) and (5) we have

$$\tilde{f}(\mu_a, x)(y) = \sup_{y=f(a,x)} \mu_a(a) = \sup_{B(x,y)} g^{(-1)}(\|\hat{D}^{-1}(\hat{a} - \hat{\alpha})\|_p,$$

where $B(x,y) = \{\hat{a} \in \mathbb{R}^k : \sum_{i=1}^k a_i h_i(x) = y - \sum_{i=k+1}^r \alpha_i h_i(x)\}$.
If $\hat{h}(x) = 0$ then $B(x,0) = \mathbb{R}^k$ and $B(x,y) = \emptyset$ if $y \neq 0$, so in the first case the supremum is reached at $\hat{a} = \hat{\alpha}$, consequently $\tilde{f}(\mu_a, x)(0) = 1$; and in the second case by definition $\tilde{f}(\mu_a, x)(y) = 0$.
If $h(x) \neq 0$ then $B(x,y) = \{\hat{a} \in \mathbb{R}^k : \sum_{i=1}^k \frac{a_i - \alpha_i}{d_i} \cdot \frac{d_i h_i(x)}{y - f(\alpha,x)} = 1\}$. Since the space of linear functionals on ℓ_p^k -space equipped with the weighted norm $\|b\|_{pW^{-1}} = \|W^{-1}b\|_p$ is isometric to the ℓ_q^k-space equipped with the weighted norm $\|z\|_{qW} = \|Wz\|_q$ and the norm of a linear functional $\sum_{i=1}^k b_i z_i$ is reciprocal of the distance between the origin and the hyperspace $\sum_{i=1}^k b_i z_i = 1$, we obtain that

$$\tilde{f}(\mu_a, x)(y) = g^{(-1)}\big(\inf_{B(x,y)} \|\hat{D}^{-1}(\hat{a} - \hat{\alpha})\|_p\big) = g^{(-1)}\big(|y - f(\alpha, x)| / \|\hat{D}\hat{h}(x)\|_q\big). \qquad \blacksquare$$

3.2. Valued equality and inequality relations between fuzzy sets

Let R be an unfuzzy relation on X. The T_{gp}-fuzzification $\tilde{R}$ of R on $\mathcal{F}(X)$ is a fuzzy set on $\mathcal{F}(X) \times_T \mathcal{F}(X)$:

$$\tilde{R}(\mu, \nu) = \sup_{xRy} T_p(\mu(x), \nu(y)).$$

Definition 3.2.1. Let $\mu_1 = (\alpha_1, d_1)$, $\mu_2 = (\alpha_2, d_2)$ be quasitriangular fuzzy numbers generated by g and let $f(a) = f(a_1, a_2) = a_1 - a_2$. Let $\tilde{f}(\mu)(y)$ be the *(g,p)*-fuzzification of f with the *(g,p)*-fuzzy vector $\mu = \mu_1 \times \mu_2 = (\alpha, d), \alpha = (\alpha_1, \alpha_2), d = (d_1, d_2)$. We say that

$$v(\mu_1 R \mu_2) = \tilde{R}(\tilde{f}(\mu), \chi_{\{0\}})$$

is a *(g,p)*-valued R-relation between μ_1 and μ_2.

Lemma 3.2.2.

$$v(\mu_1 = \mu_2) = \begin{cases} g^{(-1)}(|\alpha_1 - \alpha_2| / \|d\|_q), & \text{if } d \neq 0 \\ \chi_{\{u: u=0\}}(\alpha_1 - \alpha_2), & \text{if } d = 0 \end{cases},$$

$$v(\mu_1 \leq \mu_2) = \begin{cases} g^{(-1)}(\max(0, (\alpha_1 - \alpha_2)/\|d\|_q)), & \text{if } d \neq 0 \\ \chi_{\{u: u \leq 0\}}(\alpha_1 - \alpha_2), & \text{if } d = 0 \end{cases}.$$

Proof. This immediatelly follows from the Theorem 3.1.1. with $\hat{h}(x) = (1, -1)$ and the Definition 3.2.

$\blacksquare$

Theorem 3.2.3. *Let the equality* $f(a,x) = \sum_{j=1}^{r} a_j h_j(x) = a_0$ *be fuzzified by the (g,p)-fuzzy vector* $\overline{\mu}_a = \mu_0 \times \mu_a = (\overline{\alpha}, \overline{d}) \in \mathcal{F}_g^{r+1}$, *where* $\overline{\alpha} = (\alpha_0, \alpha_1, \ldots, \alpha_r)$, $\overline{d} = (d_0, d_1, \ldots, d_r)$, $d_0 \geq 0$, $d_j > 0$ *for* $j = 1, \ldots, k$, $d_j = 0$ *for* $j = k+1, \ldots, r$. *Then the (g,p)-valued equality relation* $v(\tilde{f}(\mu_a, x) = \mu_0)$ *is a fuzzy set* $\sigma \in \mathcal{F}(\mathbb{R}^n)$ *with the membership function:*

$$\sigma(x) = \begin{cases} g^{(-1)}\left(|f(\alpha,x) - \alpha_0| / \|\overline{D}\overline{h}(x)\|_q\right), & \text{if } \overline{D}\overline{h}(x) \neq 0 \\ \chi_{\{x: f(\alpha,x) - \alpha_0 = 0\}}, & \text{otherwise} \end{cases},$$

where $\overline{D} = \operatorname{diag}(d_0, d_1, \ldots, d_r)$, $\overline{h}(x) = (-1, h_1(x), \ldots, h_r(x))$.

Proof. It is trivial from the Theorem 3.1.1. and the first statement of the Lemma 3.2.2. ■

Theorem 3.2.4. *Let the inequality* $f(a,x) = \sum_{j=1}^{r} a_j h_j(x) \leq a_0$ *be fuzzified by the (g,p)-fuzzy vector* $\overline{\mu}_a = \mu_0 \times \mu_a = (\overline{\alpha}, \overline{d}) \in \mathcal{F}_g^{r+1}$, *where* $\overline{\alpha} = (\alpha_0, \alpha_1, \ldots, \alpha_r)$, $\overline{d} = (d_0, d_1, \ldots, d_r)$, $d_0 \geq 0$, $d_j > 0$ *for* $j = 1, \ldots, k$, $d_j = 0$ *for* $j = k+1, \ldots, r$. *Then the (g,p)-valued inequality relation* $v(\tilde{f}(\mu_a, x) \leq \mu_0)$ *is a fuzzy set* $\sigma \in \mathcal{F}(\mathbb{R}^n)$ *with the membership function:*

$$\sigma(x) = \begin{cases} g^{(-1)}\left(\max(0,\ f(\alpha,x) - \alpha_0) / \|\overline{D}\overline{h}(x)\|_q\right), & \text{if } \overline{D}\overline{h}(x) \neq 0 \\ \chi_{\{x: f(\alpha,x) - \alpha_0 \leq 0\}}(x), & \text{otherwise} \end{cases},$$

where $\overline{D} = \operatorname{diag}(d_0, d_1, \ldots, d_r)$, $\overline{h}(x) = (-1, h_1(x), \ldots, h_r(x))$.

Proof. It follows from the Theorem 3.1.1. and the second statement of the Lemma 3.2.2. ■

4. FUZZIFICATION OF THE MATHEMATICAL PROGRAMMING PROBLEM

4.1. Fuzzification of feasible set

Let us observe that the fuzzification of a linear functional reserves the duality property of the parameter and the basic space in the following sense: if the fuzziness of the parameters is measured in the norm of the parameter space, then the fuzziness of the function value is measured in the dual space norm. This observation motivates us define the fuzzy feasible set ν_C as the T_{gq}-intersection of the *(g,p)*-valued relations of the fuzzified equalities and inequalities of (2).

Theorem 4.1.1. *Let the i-th relation in* (2) *fuzzified by the (g,p)-fuzzy vector* $\mu_i = (\overline{\alpha}_i, \overline{d}_i) \in \mathcal{F}_g^{r+1}$ *for* $i = 1, \ldots, s$, *where* $\overline{\alpha}_i = (\alpha_{i0}, \alpha_{i1}, \ldots, \alpha_{ir})$, $\overline{d}_i = (d_{i0}, d_{i1}, \ldots, d_{ir})$, $d_{ij} = \beta_i \cdot d_j$, $\beta_i > 0$, $i = 1, \ldots, s$, $j = 0, 1, \ldots, r$, $d_0 \geq 0$, $d_j > 0$ *for* $j = 1, \ldots, k$, $d_j = 0$ *for* $j = k+1, \ldots, r$. *Then the (g,p)-fuzzy feasible set is defined by the membership function*

$$\nu_C(x) = \begin{cases} g^{(-1)}\left(G(x) / \|\overline{D}\overline{h}(x)\|_q\right), & \text{if } \overline{D}\overline{h}(x) \neq 0 \\ \chi_C(x) & \text{otherwise} \end{cases},$$

where

$$G(x) = \|(\mathcal{A}h(x) - \alpha^0)_+\|_{qB},$$

$(\mathcal{A}h(x)-\alpha^0)_+$ is a vector i-th coordinate of which is

$$(f_i(\alpha_i,x)-\alpha_{i0})_+ = \begin{cases} \max\{0, f_i(\alpha_i,x)-\alpha_{i0}\} & \text{for } i=1,\ldots,m \\ |f_i(\alpha_i,x)-\alpha_{i0}| & \text{for } i=m+1,\ldots,s \end{cases},$$

$\alpha^0 = (\alpha_{10},\ldots,\alpha_{s0})$, $\|\cdot\|_{qB}$ is the weighted ℓ_q-norm with the weight matrix $B = \mathrm{diag}(\beta_1,\ldots,\beta_s)$, $\overline{D} = \mathrm{diag}(d_0,d_1,\ldots,d_r)$, $\overline{h}(x) := (-1,h_1(x),\ldots,h_r(x))$ and χ_C is the characteristic function of the set defined by (2).

Proof. If $1 \le q < \infty$ then

$$\nu_C(x) = \begin{cases} g^{(-1)}\left(\sum_{i=1}^{s}[(f_i(\alpha_i,x)-\alpha_{i0})_+^q]^{1/q}/\|\overline{D}h(x)\|_q\right), & \text{if } \overline{D}h(x) \ne 0 \\ g^{(-1)}\left(\sum_{i=1}^{s} g^q(\chi_{\{x: f_i(\alpha_i,x) R_i \alpha_i, i=1,\ldots,s\}}(x))\right), & \text{othervise} \end{cases},$$

where R_i is the i-th relation in (2).

If $q=\infty$ then

$$\nu_C(x) = \begin{cases} g^{(-1)}\left(\max_{i=1,\ldots,s}(f_i(\alpha_i,x)-\alpha_{i0})_+/(\beta_i\|\overline{D}h(x)\|_q)\right), & \text{if } \overline{D}h(x) \ne 0 \\ \min_{i=1,\ldots,s} \chi_{\{x: f_i(\alpha_i,x) R_i \alpha_{i0}, i=1,\ldots,s\}}(x) = \chi_C(x), & \text{otherwise} \end{cases}.$$

In both cases the obtained formulas are equivalent with the statement of the theorem. ∎

Let $\nu^* = \sup_{x\in \mathbb{R}^n} \nu_C(x)$ and $C_\nu^* = \{x \in \mathbb{R}^n : \nu_C(x) = \nu^*\}$.

Remarks 4.1.2.

a) $\nu^* = 1$ and $C_\nu^* \ne \emptyset$ iff (2) is not empty;

b) If $\nu^* = 0$ then there is no consistent perturbation of the constraint set (2);

c) If $0 < \nu^* < 1$ then there is no solution of (1)-(2) in classical sense;

d) $G(x)$ is a usual penalty function of the domain given by (2).

In the following we will assume that $0 < \nu^* \le 1$ and $C_\nu^* \ne \emptyset$.

4.2. A concept of optimality

Let us seek the optimal fuzzy function value as an element of $\mathcal{F}_g$.

Definition 4.2.1. A fuzzy number $\mu_{00} = (y, d_{00}) \in \mathcal{F}_g$ is a *(g,p)–fuzzy aspiration level* for the objective function with the *optimality rate* $\omega(x,y)$ if $\omega(x,y)$ is the *(g,p)*–valued $\le$ relation between the fuzzified objective function and the aspiration level.

Let the objective function be fuzzified by the *(g,p)*–fuzzy vector $\mu_0 = (\alpha_0, d_0) \in \mathcal{F}_g^r$, $\alpha_0 = (\alpha_{01},\ldots,\alpha_{0r})$, $d_0 = (d_{01},\ldots,d_{0r})$. According to the Theorem 3.2.4. the optimality rate in this case is

$$\omega(x,y) = \begin{cases} g^{(-1)}\left((f_0(\alpha_0,x)-y)_+/\|\overline{D}h(x)\|_q\right), & \text{if } \|\overline{D}h(x)\|_q \ne 0 \\ \chi_{\{x: f_0(\alpha_0,x)-y\le 0\}}(x), & \text{otherwise} \end{cases},$$

where $(f_0(\alpha_0,x)-y)_+ = \max\{0, f_0(\alpha_0,x)-y\}$.

Definition 4.2.2. The *fuzzy optimum set* $\omega^*(x,y)$ with the fuzzy aspiration level μ_{00} is the T_{gq} restriction of the fuzzy set on $X \times Y$ defined by the optimality rate to the fuzzy feasible set ν_C.

If the fuzzifying vector $\overline{\mu}_0 = \mu_{00} \times \mu_0 = (\overline{\alpha}_0, \overline{d}_0)$ is such that $\overline{\alpha}_0 = (y, \alpha_0)$, $\overline{d}_0 = (d_{00}, d_{01}, \ldots, d_{0r})$, $d_{0j} = \beta_0 \cdot d_j$, $j = 0, 1, \ldots, r$, $\beta_0 > 0$ and the vector $\overline{d} = (d_0, d_1, \ldots, d_r)$ is the same which was used in the Theorem 4.1.1 for the fuzzification of the feasible set then it is easy to verify that

$$\omega^*(x,y) = T_{gq}(\omega(x,y), \nu_C(x)) = \begin{cases} g^{(-1)}(\Phi(x,y)/\|\overline{Dh}(x)\|_q), & \text{if } \overline{Dh}(x) \neq 0 \\ \chi_{C \cap \{z : f_0(\alpha_0, z) \leq y\}}(x), & \text{otherwise} \end{cases}, \tag{8}$$

where $\Phi(x,y) = \|((f_0(\alpha_0, x) - y)_+ / \beta_0,\ G(x))\|_q$.

Definition 4.2.3. The *(g,p)*–fuzzy aspiration level $\mu_0^* = (y^*, d_{00})$ is *optimal* for the fuzzified mathematical programming problem if y^* is the minimal root of the equation

$$\psi(y) = \sup_{x \in \mathbb{R}^n} \omega^*(x,y) = \nu^*$$

in the following sense: $\psi(y^*) = \nu^*$, $\psi(y) < \nu^*$ for every $y < y^*$ and $\limsup_{y \to -\infty} \psi(y) < \nu^*$. If $\limsup_{y \to -\infty} \psi(y) = \nu^*$ then $y^* = -\infty$ and we say that the fuzzified objective function is unbounded from below on the fuzzy feasible set. $x^* = x(y^*) \in \mathbb{R}^n$ realizes the optimal aspiration level if $\psi(y^*) = \omega^*(x^*, y^*)$.

Remark that due to (8) the optimal aspiration level given by the Definition 4.2.3. for the crisp mathematical programming coincides with the classical optimum. Therefore, in the sequel we deal only with such problems, which are everywhere proper fuzzified problems. Moreover, besides the convexity property of (1)-(2) we assume that $\|\overline{Dh}(x)\|_q$ is also convex.

Let us suppose that $d_0 > 0$, consequently $d_{i0} > 0$, $i = 0, 1, \ldots, s$. In this case the following equivalent formulations of the optimal *(g,p)*-aspiration level hold true, which will be very useful for the numerical method construction.

Theorem 4.2.4. *$\mu_0 = (y^*, d_{00})$ is an optimal aspiration level for the (g,p)-fuzzified version of* (1)-(2) *iff y^* is the minimal root of the equation*

$$\varphi(y) = \inf_{x \in \mathbb{R}^n} \Phi(x,y)/\|\overline{Dh}(x)\|_q = g(\nu^*) \tag{9}$$

in the following sense: $\varphi(y^*) = g(\nu^*)$, $\varphi(y) > g(\nu^*)$ for every $y < y^*$ and $\liminf_{y \to -\infty} \varphi(y) > g(\nu^*)$ or $\liminf_{y \to -\infty} \varphi(y) = g(\nu^*)$ and then $y^* = -\infty$. Moreover, if the support set $\operatorname{supp} \nu_C$ of the fuzzy feasible set is compact then the aspiration level $\mu_0 = (y^*, d_{00})$ is optimal iff y^* is the minimal root of the equation

$$F^*(y) = \inf_{x \in \mathbb{R}^n} F(x,y) = 0, \tag{10}$$

where $F(x,y) = \Phi(x,y) - g(\nu^*)\|\overline{Dh}(x)\|_q$ and the minimal root is understood as in (9).

Proof. The first statement immediatelly follows from the monotonity of the function g. In the second case it is easy to verify that both $\varphi(y)$ and $F^*(y)$ are monoton (generally not strictly) decreasing functions. Otherwise, the minimization problem (9) and (10) are equivalent with their restrictions to the compact supp ν_C, therefore there are points $x(y)$ and $x^*(y)$ contained in supp ν_C, which realize the infimum in (9) and (10), respectively.

Let y^* be the minimal root of (9) and let it be finite. Then

$$0 = \Phi(x(y^*), y^*) - g(\nu^*)\|\overline{Dh}(x(y^*))\|_q \geq \Phi(x^*(y^*), y^*) - g(\nu^*)\|\overline{Dh}(x^*(y^*))\|_q ,$$

consequently $\Phi(x^*(y^*), y^*)/\|\overline{Dh}(x^*(y^*))\|_q \leq g(\nu^*)$. Otherwise, from the construction of $\Phi(x, y)$ follows that $\Phi(x^*(y^*), y^*)/\|\overline{Dh}(x^*(y^*))\|_q \geq g(\nu^*)$. So, y^* is a root of (10). Conversely, if y^* is the minimal root of (10) and it is finite, then running the previous argument in reverse one obtains that y^* is a root of (9). It means that if there exist finite minimal roots of (9) and (10), than they coincide, moreover if one of the equations (9) and (10) has infinite minimal root, so does the other one, too. ■

The following theorem shows the connection between the classical and the fuzzy optimum.

Theorem 4.2.5. *Assume that* $\nu^* = 1,\ C_\nu^* \neq \emptyset, \text{supp}\,\nu_C$ *is compact and the Lagrangian function of the classical mathematical programming problem* (1)-(2) *has a saddle point. Then* y^* *is optimal for the classical problem* (1)-(2).

Proof. In this case $g(\nu^*) = 0$, so the method given by (1) to find the aspiration level coincides with the parametrical embedding method given in Vasil'ev (1978) to find the optimal objective function value of (1)-(2) for which the given assumptions garantee the convergence conditions. So, the statement of this theorem is equivalent to the statement proved in the mention paper. ■

4.3. A numerical method

The numerical problem to find the optimal aspiration level either by (9) or (10) is not very simple because of the complicate structure of the functions $\varphi(y)$ and $F^*(y)$. However, the Theorem 4.2.3. motivates to costruct an iterative method, which combines the two method only in limit.

Let denote $P(x,t) := \begin{cases} G^q(x) - t^q\|\overline{Dh}(x)\|_q^q, & \text{if } 1 \leq q < \infty \\ G(x) - t\|\overline{Dh}(x)\|_\infty, & \text{if } q = \infty \end{cases}$

Algorithm 4.3.1.

Initialization: Choose $t_0 \in [g(\nu^*), g(0)]$.

Iteration:

Step 1: Let x_k be a solution of the minimization problem

$$P(x, t_k) \longrightarrow \inf_{x \in \mathbb{R}^n}$$

Step 2: Compute

$$y_k = \begin{cases} f_0(\alpha_0, x_k) + \beta_0 P(x_k, t_k)^{1/q}, & \text{if } 1 \leq q < \infty \\ f_0(\alpha_0, x) + \beta_0 P(x_k, t_k) & \text{if } q = \infty \end{cases} \tag{11}$$

Step 3: Change the parameter

$$t_{k+1} = G(x_k)/\|\overline{Dh}(x_k)\|_q .$$

Step 4: If $t_{k+1} = g(\nu^*)$ then the iteration is completed, otherwise the computation is continued with the Step 1.

Theorem 4.3.2. *Let* supp ν_C *be compact and let the function* $G(x) - g(0)\|\overline{Dh}(x)\|_q > -\infty$ *everywhere on* supp ν_C. *Then the sequences* $\{x_k\},\{t_k\}$ *and* $\{y_k\}$, *defined in the Algorithm 4.3.1. are convergent and* $t_k \to g(\nu^*)$, $y_k \to y^*$, $x_k \to x^*$, *where* y^* *is the center of the optimal aspiration level and* x^* *realizes it.*

Proof. The algorithm, without the second step, represents the well-known Dinkelbach algorithm (Dinkelbach, 1967) for the solution of the nonlinear fractional programming problem

$$\inf_{x \in \text{supp } \nu_C} G(x)/\|\overline{Dh}(x)\|_q .$$

Due to the assumptions of the theorem the conditions of the convergence theorem of this methods are satisfied, therefore the sequence $\{t_k\}$ is monoton decreasing and $t_k \to g(\nu^*)$, furthermore there exists $x^* \in$ supp ν_C such that $x_k \to x^*$. Since $P(x_k, t_k) \leq 0$, therefore $y_k \leq f_0(\alpha, x_k)$, so y_k is the minimal root of the equation

$$\Phi(x_k, y) - t_k\|\overline{Dh}(x_k)\|_q = 0 . \tag{12}$$

Turning in (12) to the limit one obtains that

$$\Phi(x^*, y) - g(\nu^*)\|\overline{Dh}(x^*)\|_q = 0, \tag{13}$$

the minimal root of which is the limit point of (11). However, from(13) follows that

$$\Phi(x^*, y^*)/\|\overline{Dh}(x^*)\|_q = \inf_{x \in \mathbb{R}^n} \Phi(x, y)/\|\overline{Dh}(x)\|_q ,$$

therefore y^* is optimal and x^* realizes this optimum.

5. CONCLUSION

The given optimum concept is consistent with the classical one for the crisp mathematical programming problem, which linearly depends on the parameters, but it may be used in the case also when the mathematical programming problem has no solution because of the feasible set with strict bounds is empty, but in reality this bound are violable.

REFERENCES

Dinkelbach W. (1967) On nonlinear fractional programming. Manag.Sci., 13, 492-498.

Kovács M. (1988) On the g-fuzzy linear systems, BUSEFAL 37, 69-77.

Kovács M. (1990) On the solution of fuzzified linear systems. In: V.M.Repin,V.A.Morozov (Eds.) Cooperative works of SRCC of MU and CC of BU. (in Russian), Moscow State Univesity, 12-23.

Kovács M. (1991) Stable embedding of ill-posed linear equality and inequality systems into fuzzified systems. Fuzzy Sets and Systems (to appear)

Vasil'ev F.P. (1978) On the parametrical embedding method (in Russian) Vestnik Moskov. Univ. Ser.15. No.3. 24-32.

SOME PROPERTIES OF POSSIBILISTIC LINEAR EQUALITY SYSTEMS WITH WEAKLY NONINTERACTIVE FUZZY NUMBERS

Robert Fullér
Lehrstuhl für Unternehmensforschung, RWTH Aachen
Templergraben 64, W-5100 Aachen, Germany

Abstract Possibilistic linear equality systems (PLES) are linear equality systems with fuzzy parameters and crisp variables. We study the problem: For a given PLES with weakly noninteractive fuzzy number parameters, how does the solution change as the parameters are varied?

Keywords Triangular norm, weakly noninteractive fuzzy numbers, possiblity, extension principle, triangular fuzzy number

1. INTRODUCTION

We consider the following PLES with weakly noninteractive fuzzy number parameters of symmetric triangular form

$$\tilde{a}_i x = \tilde{b}_i, \quad i=1,\dots,m \tag{1}$$

where $x=(x_1,\dots,x_n)$ is a vector of crisp variables, $\tilde{b}_i$ is a fuzzy number, $\tilde{a}_i=(\tilde{a}_{i1},\dots,\tilde{a}_{in})$ is a vector of weakly noninteractive fuzzy numbers $\tilde{a}_{ij}$, the operations addition and multiplication by a real number of fuzzy numbers are defined via a sup-t-norm convolution and the equations are understood in a possibilistic sense.

In this paper we shall estimate how does the solution of PLES (1) change as the fuzzy parameters are varied.

2. PRELIMINARIES

A fuzzy set $\tilde{a}$ of the real line **R** with a normal upper semicontinuous and finite-support membership function is called a fuzzy number. The family of all fuzzy numbers we shall denote by **F**.

A fuzzy number $\tilde{a}$ of the real line **R** given by the membership function $\tilde{a}(t)=|a-t|s^{-1}$ if $|a-t|<s$ and $\tilde{a}(t)=0$ otherwise, is called

a symmetric triangular fuzzy number with center $a \in \mathbf{R}$ and spread $2s>0$ and we shall refer to it by the pair (a,s).

A function $T:[0,1]\times[0,1]\rightarrow[0,1]$ is said to be a triangular norm or t-norm for short (Schweizer and Sklar, 1961) if T is associative, symmetric, monotone and $T(a,1)=a$, for each $a\in[0,1]$.

The fuzzy numbers $\tilde{a}$, $\tilde{b}$ are said to be T-weakly noninteractive (Dubois and Prade, 1981) if there exists a t-norm T such that their joint possibility distribution $(\tilde{a},\tilde{b})$ is defined as

$$(\tilde{a},\tilde{b})(z)=T(\tilde{a}(z_1),\tilde{b}(z_2)), \text{ for each } z=(z_1,z_2)\in\mathbf{R}^2.$$

If $f:\mathbf{R}^2\rightarrow\mathbf{R}$ is a function, $\tilde{a}$ and $\tilde{b}$ are T-weakly noninteractive fuzzy numbers, then according to Zadeh's extension principle (Zadeh, 1965) $f(\tilde{a},\tilde{b})$ is defined as follows

$$f(\tilde{a},\tilde{b})(z) = \sup_{f(x,y)=z} T(\tilde{a}(x),\tilde{b}(y))$$

We define a metric D in $\mathbf{F}$ by the equation

$$D(\tilde{a},\tilde{b}) = \sup_{s\in[0,1]} d([\tilde{a}]^s,[\tilde{b}]^s)$$

where d is the Hausdorff metric in the nonempty compact convex subsets of $\mathbf{R}^2$, and $[\tilde{a}]^s$ denotes the s-level set of $\tilde{a}$.

Let t be a T-norm and $\tilde{a}$, $\tilde{b}$ be fuzzy numbers. The truth value of the assertion "$\tilde{a}$ is equal to $\tilde{b}$" is $Pos(\tilde{a}=\tilde{b})$ defined in Zadeh (1978) as

$$Pos(\tilde{a}=\tilde{b})=\sup_{t\in\mathbf{R}} T(\tilde{a}(t),\tilde{b}(t))=(\tilde{a}-\tilde{b})(0)$$

The following lemma can be proved directly by using the properties of metric D and a result obtained by Fullér (1990), and Fullér and Keresztfalvi (1990).

Lemma 1. Let T be a t-norm, $x,y\in\mathbf{R}$, $\tilde{a}=(a,s)$, $\tilde{b}=(b,s)$, $\tilde{c}=(c,s)$ and $\tilde{d}=(d,s)$ be fuzzy numbers of a symmetric triangular form. Then,

(i) $D(x\tilde{a}+y\tilde{b},\ x\tilde{c}+y\tilde{d}) \leq |x||a-c|+|y||b-d|$

(ii) $|(x\tilde{a}+y\tilde{b})(t)-(x\tilde{c}+y\tilde{d})(t)| \leq 2s^{-1} \max\{|a-c|,|b-d|\}$, $t\in\mathbf{R}$.

3. SOME PROPERTIES OF PLES'S

Consider the following PLES's

$$\tilde{a}_i x = \tilde{b}_i, \quad i=1,\dots,m \tag{2}$$

$$\tilde{c}_i x = \tilde{d}_i, \quad i=1,\dots,m \tag{3}$$

where $\tilde{a}_i=(\tilde{a}_{ij})$, $\tilde{c}_i=(\tilde{c}_{ij})$ are vectors of T-weakly noninteractive fuzzy numbers, and $\tilde{b}_i, \tilde{d}_i$ are fuzzy numbers.

We denote by $P_i(x)$, $Q_i(x)$ the degree of satisfaction of the i-th equation at the point $x \in R^n$ in the PLES's (2) and (3), respectively, i.e.

$$P_i(x)=\text{Pos}(\tilde{a}_i x=\tilde{b}_i), \quad Q_i(x)=\text{Pos}(\tilde{c}_i x=\tilde{d}_i)$$

where Pos denotes the T-norm-based possibility.
Then a solution of the PLES's (2) and (3) is defined by Bellman and Zadeh (1972) as

$$P(x)=\min_{i=1,m} P_i(x), \quad Q(x)=\min_{i=1,m} Q_i(x)$$

In the next theorem we shall give an estimation for the sensitivity of a solution of the PLES's to the variation of fuzzy number parameters.

Theorem 1. Let T be a t-norm $\tilde{b}_i=(b_i,s)$, $\tilde{d}_i=(d_i,s)$ and let $\tilde{a}_{ij}=(a_{ij},s)$, $j=1,\dots,n$ and $\tilde{c}_{ij}=(c_{ij},s)$, $j=1,\dots,n$ be T-weakly noniteractive fuzzy numbers. Then

$$|P(x)-Q(x)| \leqq ns^{-1} \max_{i,j}\{|a_{ij}-c_{ij}|, |b_i-d_i|\} \tag{4}$$

The proof of this theorem follows from Lemma 1.

Illustration. Let $T(u,v)=T_W(u,v)$, where T_W denotes the weak t-norm defined by $T_W(u,v)=u$ if $v=1$, $T_W(u,v)=v$ if $u=1$ and $T_W(u,v)=0$ otherwise. Then with the notation of Theorem 1 we have

$$P(x) = 1 - \frac{|Ax-b|_\infty}{s\max\{|x|_\infty, 1\}}, \quad Q(x) = 1 - \frac{|Cx-d|_\infty}{s\max\{|x|_\infty, 1\}},$$

so that

$$|P(x)-Q(x)| \leq s^{-1}(\|A-C\|+|b-d|_\infty),$$

where $|x|_\infty = |x_1|+\ldots+|x_n|$.

Remark 1. From (4) it follows that small changes in the membership function of fuzzy parameters of the PLES's (2) and (3) may cause only a small variation in the solution, i.e. unlike for crisp linear equality systems, the PLES's always belong to the family of well-posed problems.

Remark 2. Similar results have been obtain for the PLES's with noninteractive fuzzy number parameters by Fullér (1990) and Kovács (1988).

REFERENCES

Bellman R.E.,and Zadeh L.A. (1970) Decision-making in a fuzzy environment, Management Sci., **17** B141-B154.

Dubois D.,and Prade H. (1981) Additions of interactive fuzzy numbers, IEEE Trans. on Aut. Cont. **26** 926-936.

Fullér R. (1990) On Stability in Possibilistic Linear Equality Systems with Lipschitzian Fuzzy Numbers, Fuzzy Sets and Systems, 34, No.3 347-353.

Fullér R. (1989) On Stability in Fuzzy Linear Programming Problems, Fuzzy Sets and Systems, 3(30) 339-344.

Fullér R. (1990) On Product-Sum of Triangular Fuzzy Numbers, Fuzzy Sets and Systems (to appear).

Fullér R. and Keresztfalvi T. (1990), On Generalization of Nguyen's Theorem, Fuzzy Sets and Systems, (to appear).

Kovács M. (1988) Fuzzification of ill-posed linear systems, in: D.Greenspan and P.Rózsa eds., Numerical Methods, Colloquia Mathematica Societatis Janos Bolyai No.50, North-Holland, Amsterdam, 521-532.

Schweizer B.,and Sklar A. (1961) Associative functions and statistical triangle inequalities, Publ. Math Debrecen, 169-186.

Zadeh L.A. (1965) Fuzzy Sets, Inf. and Contr., **8** 338-353.

Zadeh L.A. (1978) Fuzzy Sets as a basis for a theory of possibility, Fuzzy Sets and Systems, **1** 3-28.

FUZZY PREFERENCES IN LINEAR PROGRAMMING

Jaroslav Ramík
Research Institute for Regions and Towns Development
Cujkovova 30, 704 08 Ostrava 3, Czechoslovakia

Abstract. This paper deals with a fuzzy approach to linear programming. Inexact coefficients in the constraints and objective function are considered. Our attention is focused on the problem of ranking (comparing) fuzzy values (numbers) in the constraints of LP problem.

Keywords: linear programming, fuzzy numbers, fuzzy preferences

1. INTRODUCTION

This paper presents a practical application of the fuzzy approach to linear programming (LP). We shall deal with LP with inexact coefficients, which is crucial problem in many applications. Our attention will be concentrated on a special area of ranking (comparing) fuzzy values (numbers) in the constraints of LP problem. In a certain sense this paper is a following of Ramik (1986) and Dubois (1983).

2. MOTIVATION - A CRISP CASE

The problem of LP with inexact coefficients may be stated as follows:

maximize

$$c^T x,\ c \in \mathcal{C} \subset E_n, \tag{1}$$

subject to

$$A\,x \le b,\quad A \in \mathcal{A} \subset E_{mxn},\ b \in \mathcal{B} \subset E_m, \tag{2}$$
$$x \ge 0,$$

where the coefficients A , b , c are taken from the corresponding sets $\mathcal{A}$, $\mathcal{B}$ and $\mathcal{C}$, x is a crisp decision variable vector. If $\mathcal{A}$, $\mathcal{B}$, $\mathcal{C}$ contain just one element, then we have a classical LP problem. *A feasible solution* is defined as a set

$$X = \{x \in E_n^+ \ ;\ A\,x \le b\}.$$

If $\mathcal{A}$, $\mathcal{B}$ or $\mathcal{C}$ has more than one element, then there arises a problem of how to define or understand a corresponding feasible solution set. In operations research we meet four types of "feasible" sets:

$$X_1 = \{x \in E_n^+ \ ;\ \forall A \in \mathcal{A},\ \forall b \in \mathcal{B}\ :\ A\,x \le b\},$$

$$X_2 = \{x \in E_n^+ \ ;\ \forall A \in \mathcal{A},\ \exists b \in \mathcal{B}\ :\ A\,x \le b\},$$

$$X_3 = \{x \in E_n^+ \ ;\ \exists A \in \mathcal{A},\ \forall b \in \mathcal{B}\ :\ A\,x \le b\},$$

$$X_4 = \{x \in E_n^+ \ ;\ \exists A \in \mathcal{A},\ \exists b \in \mathcal{B}\ :\ A\,x \le b\}.$$

Note that we have all four combinations of general and existence quantifiers in the above sets, X_i, $i=1,2,3,4$. This fact is not artificial, all four types of problems are practical ones.

The following simple proposition is presented without proof.

Proposition 1. It holds

$$X_1 \subseteq X_2 \text{ and } X_3 \subseteq X_4$$

and moreover, if $\mathcal{B}$ contains just one element, then

$$X_1 = X_2 \text{ and } X_3 = X_4.$$

The following proposition is a kind of representation theorem. Here max $\mathcal{A}$ (max $\mathcal{B}$), min $\mathcal{A}$ (min $\mathcal{B}$) denote maximal (minimal) subsets of the corresponding sets $\mathcal{A}$ ($\mathcal{B}$) induced by the usual ordering.

<u>Proposition 2.</u> It holds

$$X_1 = \{x \in E_n^+ \ ; \ \forall\, A \in \max \mathcal{A},\ \forall\, b \in \min \mathcal{B} : A x \le b\},$$

$$X_2 = \{x \in E_n^+ \ ; \ \forall\, A \in \max \mathcal{A},\ \exists\, b \in \max \mathcal{B} : A x \le b\},$$

$$X_3 = \{x \in E_n^+ \ ; \ \exists\, A \in \min \mathcal{A},\ \forall\, b \in \min \mathcal{B} : A x \le b\},$$

$$X_4 = \{x \in E_n^+ \ ; \ \exists\, A \in \min \mathcal{A},\ \exists\, b \in \max \mathcal{B} : A x \le b\}.$$

The situation is substantially simplified if $\mathcal{A}$, $\mathcal{B}$ are matrix intervals. Then the maximal element is unique.

<u>Example 1.</u> Let $A = \{a_{ij}\}$, $\underline{A} = \{\underline{a}_{ij}\}$, $\bar{A} = \{\bar{a}_{ij}\}$ be matrices, $b = \{b_i\}$, $\underline{b} = \{\underline{b}_i\}$, $\bar{b} = \{\bar{b}_j\}$ be given vectors; and

$$\mathcal{A} = \{A \in E_{m\times n},\ \underline{a}_{ij} \le a_{ij} \le \bar{a}_{ij},\ i=1,..,m,\ j=1,..,n\},$$

$$\mathcal{B} = \{b \in E_m \ ; \ \underline{b}_i \le b_i \le \bar{b}_i,\ i=1,...,m\},$$

be matrix and vector intervals, respectively. Then

$$\min \mathcal{A} = \underline{A},\ \max \mathcal{A} = \bar{A},\ \min \mathcal{B} = \underline{b},\ \max \mathcal{B} = \bar{b}$$

and

$$X_1 = \{x \in E_n^+ \ ; \ \bar{A}\, x \le \underline{b}\ \},$$

$$X_2 = \{x \in E_n^+ \ ; \ \bar{A}\, x \le \bar{b}\ \},$$

$$X_3 = \{x \in E_n^+ \ ; \ \underline{A}\, x \le \underline{b}\ \},$$

$$X_4 = \{x \in E_n^+ \ ; \ \underline{A}\, x \le \bar{b}\ \}.$$

The feasible solution sets X_k are normal convex polytopes generated by linear inequations.

If $\mathcal{C}$ contains more than one element, we have again the problem of how to define the objective function and, correspondingly, the optimal solution. Several proposals are known from the literature, as, e.g.,

$$f_o(x) = \sup_{c\in\mathcal{C}} c^T x \quad \text{"optimistic objective"}, \tag{3}$$

$$f_1(x) = \inf_{c\in\mathcal{C}} c^T x \quad \text{"pessimistic objective"}, \tag{4}$$

while, as a combination of the above two objectives we obtain the Hurwitz criterion, (Dumitru, 1986):

$$f_\lambda(x) = \lambda \inf c^T x + (1 - \lambda) \sup c^T x, \tag{5}$$

where $\lambda \in \langle 0,1 \rangle$ is *a parameter of pessimism.*

Considering the constraint sets X_k, k=1,2,3,4, and the objective function (5) we obtain four types of LP problems with uncertain parameters:

$(P_{\lambda,k})$ maximize $f_\lambda(x)$,
subject to $x \in X_k$.

Problems $(P_{\lambda,k})$ can be encountered namely in economics and operations research. Whereas $(P_{0,4})$ was considered by Danzig (1963), in his famous book as the generalized LP problem, $(P_{1,1})$ was considered in Soyster (1973, 1974). In Avenhaus (1980) it is shown that when X_k and $\mathcal{C}$ are convex polyhedra, problems $(P_{\lambda,k})$ can be solved as LP problems.

3. FUZZY CASE

Sometimes, the elements of the coefficient sets $\mathcal{A}$, $\mathcal{B}$, $\mathcal{C}$ are not equivalent to each other, some are, e.g., more possible for realization, some are less important, etc. Then a more suitable model for a coefficient is a fuzzy set, or, to be more specific, a fuzzy number.

In a *fuzzy linear programming problem* (FLP) we

maximize

$$\tilde{c}_1 x_1 \tilde{+} \ldots \tilde{+} \tilde{c}_n x_n, \tag{6}$$

subject to

$$\tilde{a}_{i1} x_1 \tilde{+} \ldots \tilde{+} \tilde{a}_{in} x_n \ \tilde{Q}^{con} \ \tilde{b}_i, \tag{7}$$

$$x_j \geq 0, i=1,..m, j=1,..,n,$$

where $\tilde{c}_j, \tilde{a}_{ij}$, and $\tilde{b}_i$ are fuzzy sets on E_1, whose membership functions are $\mu_{\tilde{c}_j}: E_1 \to \langle 0,1\rangle$, $\mu_{a_{ij}}: E_1 \to \langle 0,1\rangle$, $\mu_{\tilde{b}_i}: E_1 \to \langle 0,1\rangle$. The left-hand-sides and right-hand-sides of the constraints are compared by a fuzzy preference relation $\tilde{\varrho}^{con}: \mathcal{F}x\mathcal{F} \to \langle 0,1\rangle$ on fuzzy sets.

Denote

$$\tilde{a}_i \; \tilde{\varrho}^{con} \; \tilde{b}_i \equiv \tilde{\varrho}^{con}(\tilde{a}_i, \tilde{b}_i).$$

Here, the vector of decision variables x remains crisp. We seek to generalize the crisp LP problem with inexact coefficients, i.e. we shall try to find counterparts of the different feasible solution sets.

A fuzzy feasible solution set is understood as a fuzzy set on E_n^+. For this purpose we consider 4 constraint fuzzy preference relations:

$$\tilde{\varrho}_1(\tilde{u},\tilde{v}) = \max\{0, \sup\{\ \delta\in\langle 0,1\rangle\ ;\ 0\le\lambda\le\delta \Rightarrow \sup[\tilde{u}]_{1-\lambda} \le \inf[\tilde{v}]_{1-\lambda}\}\}, \quad (8)$$

$$\tilde{\varrho}_2(\tilde{u},\tilde{v}) = \max\{0, \sup\{\ \delta\in\langle 0,1\rangle\ ;\ 0\le\lambda\le\delta \Rightarrow \sup[\tilde{u}]_{1-\lambda} \le \sup[\tilde{v}]_{1-\lambda}\}\}, \quad (9)$$

$$\tilde{\varrho}_3(\tilde{u},\tilde{v}) = \max\{0, \sup\{\ \delta\in\langle 0,1\rangle\ ;\ 0\le\lambda\le\delta \Rightarrow \inf[\tilde{u}]_{1-\lambda} \le \inf[\tilde{v}]_{1-\lambda}\}\}, \quad (10)$$

$$\tilde{\varrho}_4(\tilde{u},\tilde{v}) = \max\{\ 0, \sup\{\ \delta\in\langle 0,1\rangle\ ;\ 0\le\lambda\le\delta \Rightarrow \inf[\tilde{u}]_{\lambda} \le \sup[\tilde{v}]_{\lambda}\}\}. \quad (11)$$

Here, $[\tilde{w}]_\tau = \{t\in E_1;\ \mu_{\tilde{w}}(t) \ge \tau\ \}$ is a <u>τ-level set of $\tilde{w}$</u>. Later, we shall derive simpler formulae for symmetric fuzzy numbers.

Fuzzy feasible solutions $\tilde{X}_k$, k=1,2,3,4 , are defined by their membership functions:

$$\mu_{\tilde{X}_k}(x) = \min_{i=1,\dots,m} \tilde{\varrho}_k^{con}(\tilde{a}_{i1}x_1 \tilde{+} \dots \tilde{+}\ \tilde{a}_{in}x_n, \tilde{b}_i). \quad (12)$$

<u>Theorem 1.</u> Let the membership functions $\mu_{\tilde{a}_i}$, $\mu_{\tilde{b}_i}$ be characteristic functions of intervals $\langle \underline{a}_{ij}, \bar{a}_{ij} \rangle$, $\langle \underline{b}_i, \bar{b}_i \rangle$, respectively. Then $\mu_{\tilde{X}_k}$ defined by (12) is a characteristic function of X_k defined in Example 1.

The proof of the above theorem is straightforward. This theorem states that our approach is a proper generalization of the crisp problem with vector intervals. In other words, adopting a fuzzy approach to the crisp problem we obtain the same result as when adopting a classical approach.

To handle the objective function with fuzzy coefficients $\tilde{c}_j$ we need a fuzzy goal $\tilde{d}$ and fuzzy preference relation $\tilde{\varrho}^{obj}$. Then the *fuzzy objective set* $\tilde{F}$ is defined by the membership function

$$\mu_{\tilde{F}}(x) = \tilde{\varrho}^{obj}(\tilde{c}_1 x_1 \tilde{+} \ldots \tilde{+} \tilde{c}_n x_n, \tilde{d}).$$

A fuzzy solution of the k-th type of the FLP problem is the couple of fuzzy sets $(\tilde{F}, \tilde{X}_k)$. Such a solution is however of no use in practice. For the purpose of decision making we need a crisp vector. This is the reason for the following two definitions.

A pareto solution of the k-th type of the FLP problem is a crisp vector $x^+ \in E_n^+$, such that the couple $(\mu_{\tilde{F}}(x^+), \mu_{\tilde{X}_k}(x^+)$ is a maximal element of the image $(\mu_{\tilde{F}}, \mu_{\tilde{X}_k})$ of E_n^+.

A *max-min solution* of the k-th type is a vector $x^m \in E_n^+$ which maximizes the minimal grade of the both components of the fuzzy solution.

4. FUZZY NUMBERS OF A SEMICOMMON SHAPE

In what follows we shall deal with a special case of fuzzy sets, namely, with symmetric fuzzy numbers generated by a special fuzzy number.

Let $\tilde{n}(.):E_1 \to \langle 0,1\rangle$ be a membership function of the fuzzy set $\tilde{n}$, such that

(i) $\tilde{n}(.)$ is quasiconcave, (convexity);

(ii) $\tilde{n}(0) = 1$, (normality);

(iii) $\tilde{n}$ is symmetric, ($\tilde{n}(x) = \tilde{n}(-x)$, $\forall x \in E_1$).

Then $\tilde{n}$ is called a *basic fuzzy number*.

Now, by $\mathcal{F}_n$ we denote the set of all fuzzy numbers generated by a shift $\bar{a} \in E_1$ and eventually by a spread $\sigma_a > 0$ of the basic fuzzy number, i.e.

$$\mathcal{F}_n = \{\tilde{a} \in \mathcal{F};\ \exists\ \bar{a} \in E_1,\ \sigma_a > 0:\ \tilde{a}(t) = \tilde{n}(\frac{t-\bar{a}}{\sigma_a}), \forall t \in E_1\}.$$

Then $\tilde{a}, \tilde{b} \in \mathcal{F}_n$ are said to be of a *semicommon shape*.

5. RANKING FUZZY NUMBERS

Let $\tilde{u}$, $\tilde{v}$ be fuzzy numbers of a semicommon shape, i.e.

$$\tilde{u}(t) = \tilde{n}(\frac{t-u}{\sigma_u}),\quad \tilde{v}(t) = \tilde{n}(\frac{t-v}{\sigma_v}).$$

Then, as it is easy to verify, fuzzy preference relations (8) - (11) can be expressed by very simple and explicit formulae :

$$\begin{aligned} \tilde{\varrho}_1(\tilde{u},\tilde{v}) &= 0 && \text{if } u > v, \\ &= 1 - \tilde{n}(\frac{u-v}{\sigma_u + \sigma_v}) && \text{if } u \le v, \end{aligned}$$

$$\tilde{\varrho}_2(\tilde{u},\tilde{v}) = 0 \qquad \text{if } u > v\,,$$
$$= 1 - \tilde{n}\Big(\frac{u - v}{\sigma_u - \sigma_v}\Big) \qquad \text{if } u \le v \text{ and } \sigma_u > \sigma_v\,,$$
$$= 1 \qquad \text{otherwise.}$$

$$\tilde{\varrho}_3(\tilde{u},\tilde{v}) = 0 \qquad \text{if } u > v\,,$$
$$= 1 - \tilde{n}\Big(\frac{u - v}{\sigma_u - \sigma_v}\Big) \qquad \text{if } u \le v \text{ and } \sigma_u < \sigma_v\,,$$
$$= 1 \qquad \text{otherwise.}$$

$$\tilde{\varrho}_4(\tilde{u},\tilde{v}) = 1 \qquad \text{if } u < v\,,$$
$$= \tilde{n}\Big(\frac{u - v}{\sigma_u + \sigma_v}\Big) \qquad \text{if } u \ge v\,.$$

The above explicit formulae are very useful from the computational point of view. In particular, if $\tilde{n}$ has a simple shape, e.g. linear or piecewise linear, then the FLP problem can be solved by classical methods (e.g., the simplex method). One can also use the commercial software packages.

From the theoretical point of view, the above fuzzy preference relations have many interesting properties, we present here only several of them.

Proposition 3. It holds

$$\tilde{\varrho}_1(\tilde{u},\tilde{v}) = 1 - \tilde{\varrho}_4(\tilde{v},\tilde{u}),$$
$$\tilde{\varrho}_1(\tilde{u},\tilde{v}) \le \min\{\tilde{\varrho}_2(\tilde{u},\tilde{v}),\tilde{\varrho}_3(\tilde{u},\tilde{v})\}\,,$$
$$\tilde{\varrho}_4(\tilde{u},\tilde{v}) \ge \max\{\tilde{\varrho}_2(\tilde{u},\tilde{v}),\tilde{\varrho}_3(\tilde{u},\tilde{v})\}\,.$$

Proposition 4. The following holds:

(i) $\tilde{\varrho}_1$ is irreflexive, antisymmetric, transitive.
(ii) $\tilde{\varrho}_2$, $\tilde{\varrho}_3$ are reflexive, antisymmetric, transitive.
(iii) $\tilde{\varrho}_4$ is reflexive and asymmetric.

The proofs of these assertions are easy, we leave them to the reader.

Let us turn to an economic interpretation of the above fuzzy preference relations. Suppose for a moment that $\tilde{u}$ represents the demand of a commodity or goods, $\tilde{v}$ represents the supply of the same commodity or goods. Then the fuzzy demand $\tilde{u}$ is $\tilde{\varrho}_1$-less then the fuzzy supply $\tilde{v}$ with the grade of possibility at least α , if and only if for any d with $\mu_{\tilde{u}}(d) \geq 1-\alpha$ and any supply s with $\mu_{\tilde{v}}(s) \geq 1-\alpha$ we have $d \leq s$,(see Fig. 1a). Even more transparent is the economic interpretation of the preference relation $\tilde{\varrho}_2$: for any demand d with $\mu_{\tilde{u}}(d) \geq 1-\alpha$, there exists a supply s with $\mu_{\tilde{v}}(s) \geq 1-\alpha$, such that $d \leq s$,(see Fig. 1b). Such an interpretation is very practical, namely in mathematical economy and operations research.

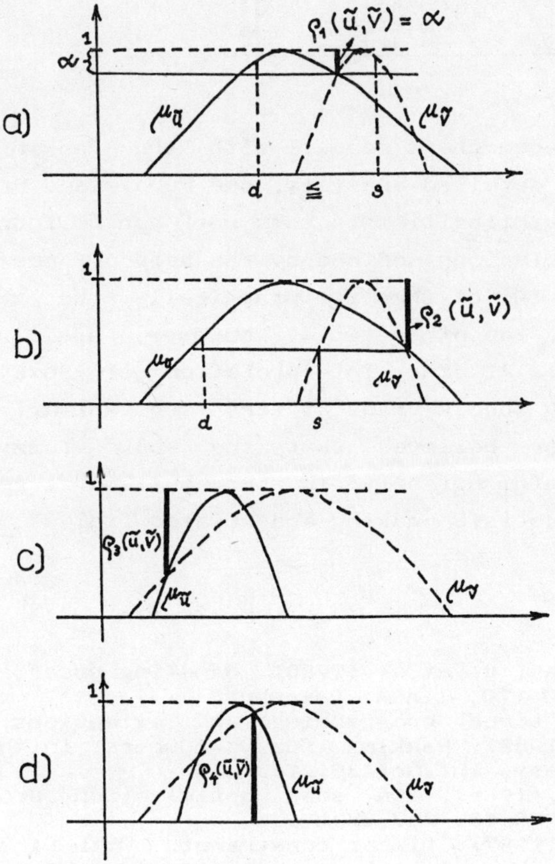

Fig. 1

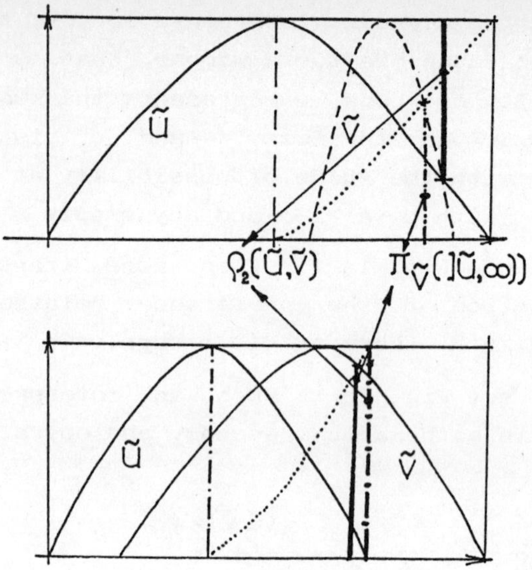

Fig. 2

6. CONCLUSION

A number of researchers dealt with the ranking of fuzzy numbers, for a more detailed analysis, see Dubois and Prade, (1983). In this paper some similarities to this work can be found: there are also 4 preference relations defined by the help of possibility and necessity measures. Two of them are practically the same as our relations $\tilde{\varrho}_1$ and $\tilde{\varrho}_4$. The other two , however, are substantially different, see Fig. 2. The interpretation of fuzzy preference relations defined by Dubois-Prade differs substantially, too. The author of this paper believes that the above fuzzy preference relations can be useful not only in LP, but more generally, in decision analysis, decision making and related fields.

References.

Avenhaus, R., Beedgen, R. et al.(1980). Handling uncertainties in a LP models, WP-80-170, IIASA, Laxenburg

Dantzig,G.B.(1963). Linear programming and extensions, Princeton

Dubois,D.,Prade,H.(1983). Ranking fuzzy numbers in the setting of possibility theory, Inf.Sci.30, 183-224

Dumitru,V., Luban,F.(1986). On some optimization problems under uncertainties, FSS 18, 257-272

Ramík,J.,Římánek,J.(1987). Linear constraints with inexact data, Tekhnicheskaya kibernetika 2, 41-48 (in russian)

Ramík,J.,Římánek,J.(1986). Inequality relation between fuzzy numbers and its use in fuzzy optimization, FSS 16, 549-564

Soyster,A.L.(1973). Convex programming with set-inclusive constr. and application to inexact LP, Oper.Res.21,1155-1157

Soyster,A.L.(1974). A duality theory for convex programming with set-inclusive constraints, Oper.Res.22, 892-898

IMPLICATION RELATIONS, EQUIVALENCE RELATIONS AND HIERARCHICAL STRUCTURE OF ATTRIBUTES IN MULTIPLE CRITERIA DECISION MAKING

Jochen Ester
Department of Automation
Chemnitz University of Technology
PF 964,0-9010 Chemnitz, Germany

Abstract.: This paper presents an efficient technique for computing implication structures, especially for Multiple Attribute Decision Making (MADM) problems. The method based on the generalized modus ponens. Implication means that there is a membership of the binary relation that a good performance level of the attribute A_i implies a good performance level of an other attribute A_j for each combination. Obviously, if such implications are found one can get some positive effects:

1. reduction of the number of relevant attributes,
2. computing a number measuring the inner contradictions between the attributes of the decision making problem,
3. obtaining a hierarchy of the attributes.

Equivalence classes arise during the computation. Basic information is given by the decision matrix. The method needs upper and lower satisfaction levels for each single criterion. Using a so-called implication threshold, given by the decision maker, one can get a crisp implication structure of crisp equivalence classes. A measuring number of the inner contradiction leads to the degree of carefulness, needed in the following MADM steps. Some methods in MADM (like, for instance, Saaty's hierarchical method) are based on a hierarchical structure of attributes. The method, presented in this paper, helps in this case, if such a hierarchy cannot be given by the decision maker.

Keywords: Multicriteria Decision Making, Fuzzy Logic, Decision Support Systems

1. INTRODUCTION

A Multiple Attribute Decision Making (MADM) problem is given by

$$\text{"max"} \{ q^k , k=1(1)K \} , \tag{1}$$
$$\{ q^k , k=1(1)K \} = Z \subset R^m , \tag{2}$$

$q^k = (q_{1k} \ldots q_{mk})^T$,

q_{ik} : value of the i-th criterion of the k-th alternative ,

where Z is a discrete set.

In many MADM cases the number m of attributes A_i (i=1(1)m) is very large. Consequently, many problems are arising for the decision maker in the evaluation process. The efficiency of Decision Support Systems (DSS) decreases with an increasing number of relevant attributes (c.f. Outranking relation methods, weighting methods, aspiration level methods and others).

Especially, much time for the questionaire procedure is needed. Therefore, it is reasonable to investigate the following problem for a concrete MADM task:

> Are there any attributes A_i such that good performance levels of them imply good performance levels of other attributes?

If such implications are found one can get some positive effects like:

1) reduction of the number of attributes,

2) defining and computing a number measuring the inner contradiction of the decision making problem,
3) obtaining a hierarchical structure of attributes.

The symbolic expression

$$A_i \to A_j = S_{ij} \quad , \tag{3}$$

will mean that attribute A_i implies attribute A_j and S_{ij} is the corresponding binary relation.

The principle problem is to find out the truth values s_{ij} for all binary relations S_{ij} $(i,j = 1(1)m)$. This paper presents an efficient technique for computing these values, based on Fuzzy Logic, especially on the generalized modus ponens.

2. PERFORMANCE LEVEL AND MEMBERSHIP

Basic information is given by the decision matrix Q with the "natural" performance levels q_{ik}.

To unify the treatment of all attributes it is necessary to compute the values μ_{ik} where μ_{ik} is the membership of the k-th alternative in the fuzzy set of all those alternatives which have good performance levels with respect to the attribute A_i. This can be done with the following piecewise linear transformation:

$$\mu_{ik} = \begin{cases} 1 & \text{for } d \geq 1 \\ d & \text{for } 0 \leq d \leq 1 \\ 0 & \text{for } d \leq 0 \end{cases} \tag{4}$$

where

$d = (q_{ik} - q_{iu})/(q_{io} - q_{iu})$,

q_{io} : upper satisfaction level,

q_{iu} : lower satisfaction level.

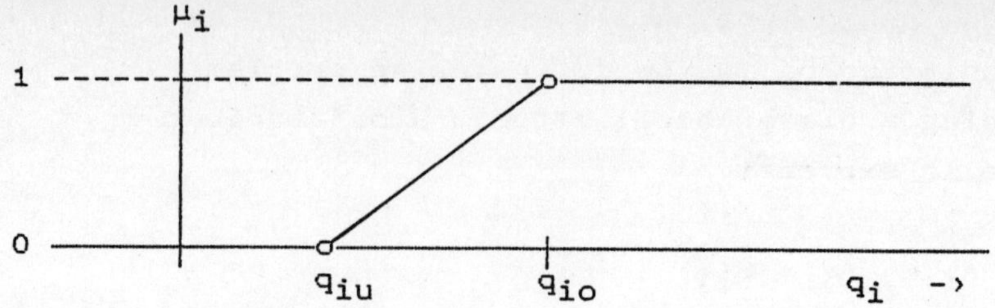

Fig.1: Membership function μ_i (maximization)

But it must be emphasized that the thresholds q_{io} and p_{iu} in general must be given by the decision maker. This is very important because the wanted truth values s_{ij} strongly depend on these thresholds. In case that the decision maker is unwilling or unable to prescribe these thresholds one can use

$$\text{(maximization)}: \quad q_{io} = \max(q_{ik}), \; q_{iu} = \min(q_{ik}),$$
$$\text{(minimization)}: \quad q_{io} = \min(q_{ik}), \; q_{iu} = \max(q_{ik}). \qquad (5)$$

Naturally, the sensitivity of performance levels, depending on other performance levels, must bee evaluated by the person concerned (decision maker).

3. COMPUTATION OF THE MEMBERSHIP s_{ij}

All implication relations between the attributes A_i can be investigated by the pairwise comparisons. According to the Theory of Fuzzy Sets, the membership s_{ij} of the relation S_{ij} is based on the membership values s_{ijk} (k=1(1)K) for every single alternative. The global implication S_{ij} for the whole problem ("at all") holds if it is fulfilled for each alternative, hence

$$s_{ij} = \text{conj}(s_{ij1},...,s_{ijK}) \qquad (6)$$

With respect to fuzzy implications the following conditions must be fulfilled (Kaufmann, 1988):

a) $\mu_i , \mu_j \in [0,1] \; , \; s_{ij} \in [0,1]$. (7.1)

b)
$$(\mu_i = 0) \,\&\, (\mu_j = 0) \rightarrow s_{ij} = 1 \,,$$
$$(\mu_i = 0) \,\&\, (\mu_j = 1) \rightarrow s_{ij} = 1 \,,$$
$$(\mu_i = 1) \,\&\, (\mu_j = 1) \rightarrow s_{ij} = 1 \,,$$
$$(\mu_i = 1) \,\&\, (\mu_j = 0) \rightarrow s_{ij} = 0 \,, \qquad (7.2)$$

c)
$$(\mu_i < \mu_i') \,\&\, (\mu_j = \mu_j') \rightarrow s_{ij} \geq s_{ij}' \,,$$
$$(\mu_i = \mu_i') \,\&\, (\mu_j < \mu_j') \rightarrow s_{ij} \leq s_{ij}' \,. \qquad (7.3)$$

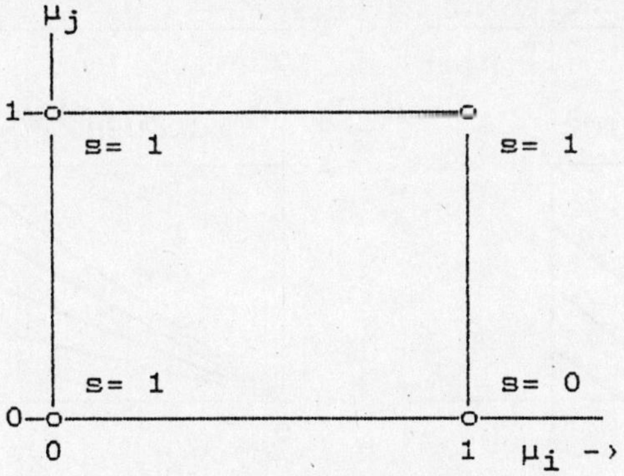

Fig.2: Main truth values of S_{ij} (s)

Operations with such membership functions where the conditions given above are fulfilled are fuzzy modifications of the so-called modus ponens.

Now, we consider four different fuzzy implications:

1. LEE:

$$s_{ijk} = \max \left(1 - \mu_{ik} \,,\, \mu_{jk} \right) \,, \qquad (8)$$

2. GOEDEL:

$$s_{ijk} = \begin{cases} 1 & \text{for } \mu_{ik} \leq \mu_{jk} \\ \mu_{jk} & \text{for } \mu_{ik} > \mu_{jk} \end{cases} \,, \qquad (9)$$

3. LUKASIEWICZ:

$$s_{ijk} = \min \left(1 \,,\, 1 - \mu_{ik} - \mu_{jk} \right) \,, \qquad (10)$$

4. GOGUEN:

$$s_{ijk} = \begin{cases} 1 & \text{for } \mu_{ik} \le \mu_{jk} \\ \mu_{ik}/\mu_{jk} & \text{for } \mu_{ik} \ge \mu_{jk} \end{cases} \tag{11}$$

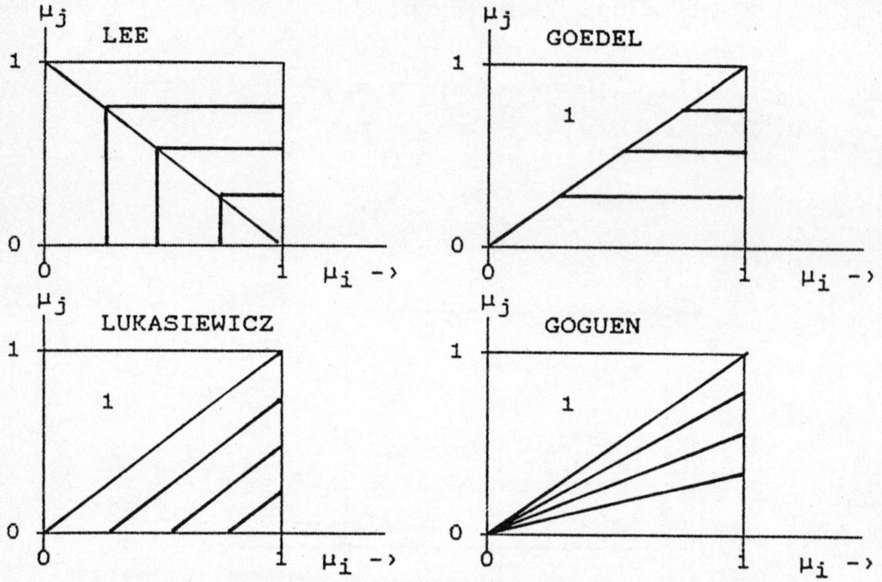

Fig.3: Different fuzzy implication rules

With the minimum operator for the intersection (6) we get

$$s_{ij} = \min\ (s_{ij}\ , \ldots,\ s_{ijK}\ )\ . \tag{12}$$

This means that the value s_{ij} is very small or zero if at least one of the alternatives with a small implication value exists. Let us carry out an imaginary experiment:

For the computation of s_{ijk} we use GOEDEL's implication. The decision maker chooses the upper thresholds using (5). For some criteria q_j the decision maker demands relatively high levels for the lower bounds q_{ju} and for other criteria q_i he demands low levels q_{iu}, respectively. Therefore, for some alternatives we get membership values μ_{jk} near or equal to zero. In this case it is easy to prove that s_{ijk} and using (12) s_{ij} get low values, too. This phenomenon, possibly for all combinations (i,j), signifies that the decision maker feels strong contradictions between some

Another important relation in the set of attributes is equivalence. Equivalence of attributes means

$$A_i <-> A_j = (A_i -> A_j) \& (A_j -> A_i) = A_{ij} = A_{ji} . \qquad (13)$$

With the minimum operator we get

$$a_{ij} = \min (s_{ij} , s_{ji}) = a_{ji} . \qquad (14)$$

Obviously, with the computation of the matrix $S = (s_{ij})$ we get all information about implications and equivalences. It must, however, be stressed that these implications and equivalences are relations entirely based on the performance of attributes in a concrete MADM problem and not at all on semantic aspects.

4. ANALYSIS OF THE IMPLICATION MATRIX S

Let us consider the implication matrix S in more detail.

	A_1	A_2	...	A_m
A_1 —>	s_{11}	s_{12}	...	s_{1m}
A_2 —>	s_{21}	s_{22}	...	s_{2m}
⋮	⋮	⋮	...	⋮
A_m —>	s_{m1}	s_{m2}	...	s_{mm}

Table 1: Implication matrix S

The elements of S are located in the closed interval [0,1]. Especially, all s_{ii} are equal to 1.

In the next step we transform S into a Boolean matrix B. This can be carried out by the use of the so-called implication theshold s, given by the decision maker

$$B = (b_{ij}) \ , \ b_{ij} = \begin{cases} 1 \text{ for } s_{ij} \geq s \\ 0 \text{ otherwise} \end{cases} \tag{15}$$

B can be illustrated as a directed graph. the nodes of which represent the attributes A_i and whose directed edges (arrows) point from implying to implied attributes. Thus, one can read off the diagramm:

a) attributes to which no arrow points cannot be omitted and replaced (substituted),

b) attributes to which arrows only point can be represented by other attributes,

c) attributes which imply each other form an equivalence class.

Transitivity of the implication relation plays a very important role. It can be proved that in the context of our concept only GOEDEL's rule is transitive

$$(b_{ik} = 1) \ \& \ (b_{kj} = 1) \rightarrow (b_{ij} = 1) \ . \tag{16}$$

It is easy to show that an increasing of s leads to graphs which are subgraphs of their predecessors.

Considering the case s=1, we get (in general) few implications only (possibly B is the unit matrix, without any implications). With decreasing s the number of implications increases and the number of equivalence classes decreases. For s=0, all attributes are equivalent. This leads to the following conclusions:

a) the chosen value of s represents the sensitivity of the decision maker (his feeling) with respect to the contradictions between the attributes or classes of attributes (equivalence classes),

b) by the gradual transition from s=1 to s=0 a multilevel hierarchy of attributes is being constructed (bottom up),

c) the distance between two thresholds of s, which charcterize an interval with stable structure of equivalence classes, can be interpreted as a number measuring the stability of this structure.

5. EXAMPLE

We shall investigate the problem of choosing a concrete technology for refining cotton fabric. For the purpose of this demonstration we have reduced the practical problem from 25 criteria to 11,and the number of alternatives is 16.

attribute	units of	max/min
1. total cost	M/1000qm	min
2. washfastness	mark 1 - 5	max
3. handle	mark 0 - 1	min
4. crease recovery	mark 1 - 5	max
5. tensile strength	N	max
6. modification of dimensional stability	%	min
7. fastness of ironing	mark 1 - 5	max
8. water vapor permeability	mg/qcm 24h	min
9. power to tear glide seam	N	max
10.power to tear out seam	N	max
11.breaking elongation	%	max

Table 2: Attributes

The values of the criteria differ in the following intervals and the thresholds are as given in Table 3.

criterion	best value = q_{io}	worst value	q_{iu}
1	188.6	387.2	400.0
2	4.7	1.0	1.0
3	0.0	0.9	1.0
4	4.7	3.2	3.0
5	599.0	488.0	450.0
6	0.9	2.4	2.5
7	5.0	5.0	5.0
8	1.0	2.2	3.0
9	97.0	76.0	70.0
10	243.0	149.0	130.0
11	13.4	8.5	7.0

Table 3: Thresholds

Using a new module of the DSS POLYP we were able to compute the following results which are shown in Fig.4 and Fig.5:
Minimizing the maximum-distance to the ideal point (with respect

to all criteria: $\mu_i = 1$, i=1(1)11), we get the rank-order of alternatives as follows (beginning with the best):

(9,11,8,13,14,1,2,10,15,3,4,16,6,5,7,12).

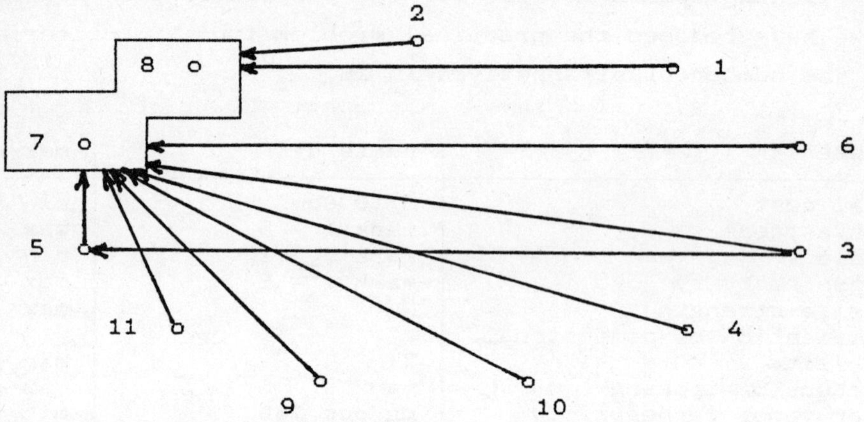

Fig.4 shows the implication structure for s=0.275
(criteria 7 and 8 are equivalent)

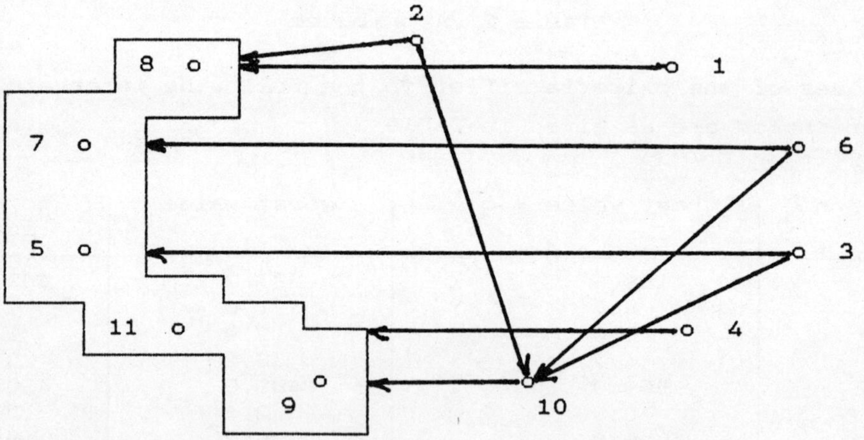

Fig.5 shows the implication structure for s=0.222
(criteria 5,7,8,9,11 are equivalent)

Now we delete the criteria 5,7 and 8 (Fig.4)! In this case minimizing the maximum-distance to the ideal point, we get the rank-order of alternatives as above (beginning with the best):

(9,11,8,13,14,1,2,10,15,3,4,16,6,5,7,12).

Deleting criteria 5,7,8,9,11 and 10 (Fig.5) the corresponding rank-order is as follows:

(9,10,11,15,8,2,13,14,1,4,3,16,6,5,7,12).

The implication structure is stable in the interval $s \in (0.45,1]$. This stable structure is characterized by the implication of the attribute "fastness of ironing" by all the other attributes. Other implications do not exist. This large interval of stability leads to the conclusion that there is a strong contradiction between all criteria except for "fastness of ironing" in this problem. Deleting this attribute is evident because its performance levels are the same for all alternatives.

Last but not least, by varying the implication threshold s, we obtain the hierarchy illustrated in Fig.6.

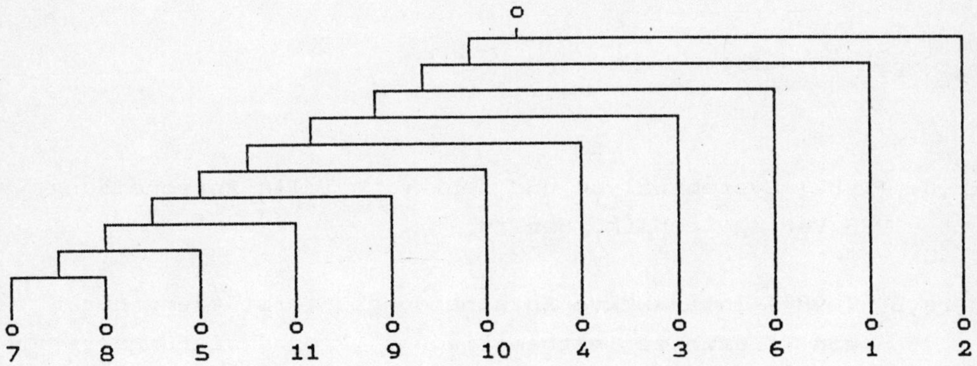

Fig.6: Hierarchy of attributes

For s=0 we get one single equivalence class only which includes all attributes (top of the hierarchy). For $0 < s \leq 0.06$ this class splits into two classes, the second of which consists of the attribute 2 only. The first class splits in the intervall $0.06 < s \leq 0.062$ into two classes, the second of which consists of the attribute 1 only, and so on. For $s > 0.45$ we find the situation that there is no equivalence class containing more than one element (attribute):

The "goodness" is split into its single features!

For example, the problem of choosing the best motorbicycle, produced in the GDR, leads to the hierarchy illustrated in Fig.10. Using this method, we can construct in a very natural way the hierarchical and logical structure of efficiency which is needed in some methods (c.f. Ester (1987)).

Attributes:
1. power,
2. max. speed,
3. fuel consumption,
4. load,
5. price,
6. taxes.

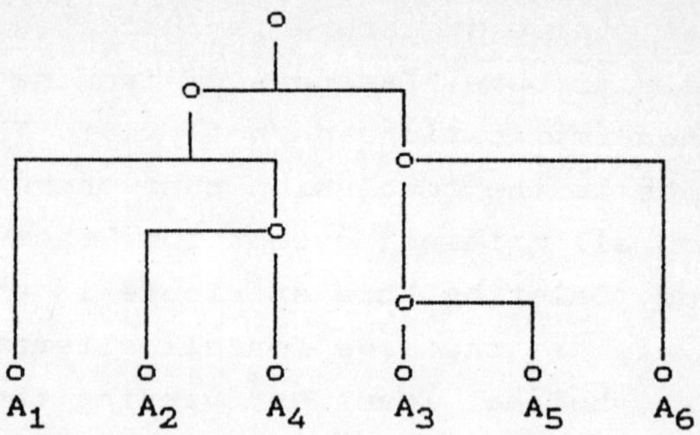

Fig.7: Hierarchy of attributes

REFERENCES

Ester,J. (1987) Systemanalyse und mehrkriterielle Entscheidung, VEB Verlag Technik. Berlin

Werners,B. (1984) Interaktive Entscheidungsunterstützung durch ein flexibles mathematisches Programmierungssystem, Minerva Publikation, München

Ester,J. (1987) A fuzzy concept of efficiency, in Jahn,J and Krabs,W.(Eds.):Recent advances and historical development of vector optimization, Springer Verlag, Berlin-Heidelberg-New York

Ester,J. (1987) Einbeziehung der Theorie der unscharfen Mengen in die Methoden der mehrkriteriellen Entscheidung, WZ der TH Ilmenau, 33(1987)6

Kaufmann,A. (1988) Logics for expert-systems, in Gupta,M.M. and Yamakawa,T.(Eds.):Fuzzy computing, North Holland

UNCERTAIN MULTIOBJECTIVE PROGRAMMING AS A GAME AGAINST NATURE

J.J. Buckley
Mathematics Department
University of Alabama at Birmingham
Birmingham, AL 35294, USA

Abstract. We consider a multiobjective linear program whose parameters have uncertain values. We first present a method of modelling the uncertainty, based on t-norms and co-t-norms, which contains possibility theory and probability theory as special cases. We then present a procedure for modelling the decision problem, which is independent of the uncertainty model, based on a game against Nature.
Keywords: mathematical programming, uncertainty modelling, games against Nature.

1 The Problem

The problem is to "solve" the multiobjective linear program

$$Max\, Z = (z_1, \ldots, z_K)\; subject\; to : Ax \leq b, x \geq 0, \tag{1}$$

where $z_k = c_{k1}x_1 + \ldots + c_{kn}x_n$ for $1 \leq k \leq K, A = [a_{ij}]$ is a mxn matrix, $b^t = (b_1, \ldots, b_m)$, and $x^t = (x_1, \ldots, x_n)$ is a vector of non-negative decision variables. For simplicity, we are considering a linear program but our results may be easily extended to non-linear problems. We have the added complication that the values of the parameters a_{ij}, b_i, and c_{kj} are uncertain.

In the next section we introduce our general method for modelling the uncertainty of the parameters a_{ij}, b_i and c_{kj}. The procedure is based on t-norms and co-t-norms and contains, as special cases, both probability and possibility. The values of the uncertain parameters will be the values of uncertain variables. These uncertain variables may be either discrete or continuous variables. The discrete case was treated in some detail in Buckley (1989) and so, in this chapter, we will assume we are working with continuous variables. The continuous case has recently been presented in Buckley (1990c) and a condensed version appears in the Appendix at the end of this chapter, with an outline of the method in the second section below.

In the third section we present our solution method. We model the decision problem (choose a "best" $x \geq 0$) as a game versus Nature. The solution technique is independent of the method of modelling the uncertainty. That is, we could be talking about stochastic programming, or possibilistic programming, or ... and the solution method is still applicable. We have argued before in Buckley (1990a) and Buckley (1990b) that a good way to model this decision problem is by modelling it as a game against Nature. What is new in this paper is that we develop the decision problem within a general framework of uncertainty modelling.

The last section contains a brief summary and our conclusions.

2 Uncertainty

Corresponding to each uncertain parameter a_{ij} there is an uncertain variable $\hat{a}_{ij} = \Omega \rightarrow R$, where Ω is some abstract space and R denotes the real numbers. A value of $\hat{a}_{ij}$ is a possible numerical value to be used for a_{ij} in the optimization problem in equation (1). Similarly, there are uncertain variables $\hat{b}_i : \Omega \rightarrow R$ and $\hat{c}_{kj} : \Omega \rightarrow R$ corresponding to uncertain parameters b_i and c_{kj}, respectively.

Each uncertain variable has a corresponding uncertainty distribution (u.d.). Let $A_{ij} : R \rightarrow [0, +\infty)$, a bounded non-negative function on the real numbers, denote a u.d. for $\hat{a}_{ij}$. Similarly, let $B_i : R \rightarrow [0, +\infty)$ and $C_{kj} : R \rightarrow [0, +\infty)$ be bounded non-negative functions on R, which are u.d.'s for $\hat{b}_i$ and $\hat{c}_{kj}$, respectively. We will now need to outline our general theory of uncertain variables and uncertainty distributions, and then we may return to the uncertain parameters in the programming problem.

Let T be any t-norm and C any co-t-norm. First, T and C are extended, using their associativity, to be functions of m arguments ($m \geq 2$). Then T and C will map $[0,1]^m$ into $[0,1]$. If $B[R^m_+], m \geq 1$, represents all bounded, non-negative, functions defined on R^m, then S_o will denote some special subset of $B[R^m_+]$. Given C and S_o we next extend C to be a mapping from S_o into $[0, +\infty)$. Some of the details are in the Appendix with complete details in Buckley (1990c). We therefore have $C(f) \geq 0$ for each $f \epsilon S_o$. Let $\mathcal{A}$ be a collection of subsets of R^m so that $f\chi_E \epsilon S_o$ for $f \epsilon S_o$ and E in $\mathcal{A}$. We therefore have $C(f\chi_E) \geq 0$, for f in S_o and E in $\mathcal{A}$, which equals $C(f(x)|x \epsilon E)$ because C will ignore all the zero values in the function $f\chi_E$.

If $F : \Omega \rightarrow R^m$ is an uncertain variable, then it will have an uncertainty distribution f in S_o. We write

$$Put\,[F \epsilon E] = C(f(x)|x \epsilon E), \tag{2}$$

for E in $\mathcal{A}$, where we use the term "Put" as an abbreviation for the word "putting." Equation (2) measures our ability to put the values of F into the set E.

We may choose S_o, C, and $\mathcal{A}$ to have Put = Poss for possibility or Put = Prob for probability. For example, if f in a normalized fuzzy subset of R^m, C = supremum the extension of C = max, and $\mathcal{A}$ is all subsets of R^m, then Put = Poss and we have

$$Poss\,[F \epsilon E] = sup(f(x)|x \epsilon E). \tag{3}$$

Here F is called a fuzzy variable. To get probability, let f be a probability density function (area, or volume, under the graph is one), C = integration an extension of $C = LOR$, and $\mathcal{A}$ = measurable subsets of R^m, then Put = Prob and

$$Prob\,[F \epsilon E] = \int_E f(x)dx. \tag{4}$$

Now F is called a random variable.

In general, let $\theta[E] = Put[F\epsilon E]$. In any case we can choose S_o, C, and $\mathcal{A}$ so that θ will be a generalized fuzzy measure on $\mathcal{A}$. Assuming that R^m and ϕ belong to $\mathcal{A}$, then θ is a generalized fuzzy measure if : (1) $\theta[R^m] = L > 0$; (2)$\theta[\phi] = 0$; and (3) if $E_1, E_2 \epsilon \mathcal{A}, E_1 \subset E_2$, then $\theta[E_1] \leq \theta[E_2]$. It satisfies the requirements to be a fuzzy measure in the sense of Sugeno (1974) except $\theta[R^m]$ can exceed one. The generalized fuzzy measure is used in Buckley (1990c) to further develop the theory of uncertain variables.

Now we may return to the uncertain parameters in our optimization problem. We have

$$Put\,[\hat{a}_{ij} \epsilon E] = C(A_{ij}(x) | x \epsilon E), \tag{5}$$

$$Put\,[\hat{b}_i \epsilon E] = C(B_i(x) | x \epsilon E), \tag{6}$$

and

$$Put\,[\hat{c}_{kj} \epsilon E] = C(C_{kj}(x) | x \epsilon E), \tag{7}$$

for certain choices of S_o, C, and $\mathcal{A}$. Equations (5) through (7) measure our ablility to put the values of the uncertain parameters into set E in $\mathcal{A}$.

Since we have multiple uncertain variables we will need certain joint distributions. Now T will be some continuous function mapping $[0, +\infty)^m$ into $[0, +\infty)$. T can be an extended t-norm because some t-norms automatically extend to arguments in $[0, +\infty)$. For example, T= min, T= PAND, and T= LAND are defined for arguments greater than one. Let

$$\Pi(A, b) = T(A_{ij}(a_{ij}), B_i(b_i) | 1 \leq i \leq m, 1 \leq j \leq n), \tag{8}$$

$$\Pi(c) = T(C_{kj}(c_{kj}) | 1 \leq k \leq K, 1 \leq j \leq n), \tag{9}$$

and

$$\Pi(A, b, c) = T(A_{ij}(a_{ij}), B_i(b_i), C_{kj}(c_{kj}) | 1 \leq i \leq m, 1 \leq j \leq n, 1 \leq k \leq K). \tag{10}$$

We cannot use a regular t-norm T since the A_{ij}, B_i, or C_{kj} values may exceed one. In equation (8) the value of the mapping $A_{ij} : R \rightarrow [0, +\infty)$ at $x = a_{ij} \epsilon R$ is written as $A_{ij}(a_{ij})$. Similarly we interpret $B_i(b_i)$ as the value of B_i at $x = b_i$ and $C_{kj}(c_{kj})$ as the value of C_{kj} at $x = c_{kj}$. We assume that $\Pi(A, b), \Pi(c)$, and $\Pi(A, b, c)$ all belong to S_o so they may represent joint uncertainty distributions for uncertain variables. When we can form the joint distributions as in equations (8) through (10), using some extended t-norm, we say that the corresponding uncertain variables are non-interfering. If the uncertain variables are interfering, then we assume that the joint distributions are formed by some other procedure and we do not use equations (8) through (10).

In possibility theory, the A_{ij}, B_i, and C_{kj} are all normalized fuzzy subsets of R, T= min, and the fuzzy variables are non-interactive (Zadeh (1975), Zadeh (1978)) to obtain the joint possibility distributions as in equations (8) through (10). In probability theory, the A_{ij}, B_i, and C_{kj} are all probability density functions, T = PAND(multiplication), and we say the random variables are independent to obtain the joint distributions using equations (8) to (10). In any case, we assume to have an S_o, C, and $\mathcal{A}$ so that the joint distributions all belong to S_o.

3 Solution

We will now use the joint distributions to model the decision problem as a game versus Nature(Luce and Raiffa (1957)). We view the multiobjective programming problem in equation (1) as an infinite number of problems each having a different uncertainty measured by $\Pi(A, b, c)$.

A game against Nature has three ingredients: (1) the actions that may be chosen by the decision maker; (2) the states of Nature; and (3) the payoff to the decision maker given his action and the resulting state of Nature. The actions are obviously the values $x \geq 0$ of the decision variables. The states of Nature are the multiobjective programming problems given in equation (1). We obtain different states of Nature by assigning different values to the uncertain parameters a_{ij}, b_i, and c_{kj}. Each state of Nature has a known uncertainty $\Pi(A, b, c)$. The payoff to the decision maker is the gain (or loss) associated with choosing some value of $x \geq 0$ and having the resulting state of Nature given by specific values of A, b, and c(which determine the multiobjective programming problem). However, the payoff will not be just one number but instead a vector because of the complexity of the decision problem.

We have identified three, possibly conflicting, goals for a decision maker. The goals are: (1) make $x \geq 0$ feasible; (2) make $x \geq 0$ undominated; and (3) make Z as large as possible. We will now quantify these goals in order to completely specify the payoff vector to the game. Then we may discuss strategies on choosing a "best" $x \geq 0$.

First consider the goal of making $x \geq 0$ feasible. The decision maker does not want to choose an $x \geq 0$ so that it is usually infeasible to the resulting optimization problem. For $x \geq 0$ define

$$E_1(x) = \{(A, b) | Ax \leq b\}, \tag{11}$$

and

$$g_1(x) = Put\,[x\, feasible] = C(\Pi(A, b) | (A, b) \epsilon E_1(x)). \tag{12}$$

The decision maker wishes to find an $x \geq 0$ to maximize $g_1(x)$.

Next consider the goal of making $x \geq 0$ undominated. If $v \geq 0$ is a vector in R^n, let us write $z_k(v)$ for $c_{k1}v_1 + \ldots + c_{kn}v_n, 1 \leq k \leq K$. Let $x \geq 0$ be feasible. We say x is undominated (belongs to the Pareto optimal set, is an efficient point, ...) if there is no feasible $v \geq 0$ so that $z_k(v) \geq z_k(x)$ all k and $z_k(v) > z_k(x)$ for at least one value of k. The decision maker does not want to choose an $x \geq 0$ which is usually dominated in the resulting multiobjective programming problem. For $x \geq 0$ define

$$E_2(x) = \{(A, b, c) | x\; is\; feasible\; and\; undominated\}, \tag{13}$$

and

$$g_2(x) = Put\,[x\; undominated] = C(\Pi(A, b, c) | (A, b, c) \epsilon E_2(x)). \tag{14}$$

The decision maker wishes to obtain an $x \geq 0$ to maximize $g_2(x)$. We would expect that $g_2(x) \leq g_1(x)$.

Lastly, consider the goal of making the objective function Z as large as possible. Surely, the decision maker does not want to pick an $x \geq 0$ which usually produces small values for the z_k. Let $I = (z_1^*, \ldots, z_K^*)$ be some ideal point. For example, z_k^* is usually picked so that it exceeds $z_k(x)$ for feasible $x \geq 0$ with respect to all multiobjective programs for all reasonable values of A, b, and c. That is, if $\Pi(A, b, c) > 0$, then $z_k^* > z_k(x)$ for feasible $x \geq 0$. Next let S(I) be some subset of R^K containing I. S(I) will be the target region for Z and we will say that Z is "big" whenever Z belongs to S(I). For example, choose $r_k < z_k^*, 1 \leq k \leq K$, then we could use S(I) $= \{v | v_k \geq r_k, 1 \leq k \leq K\}$. For $x \geq 0$ define

$$E_3(x) = \{c | z_k(x) \epsilon S(I), 1 \leq k \leq K\}, \tag{15}$$

and

$$g_3(x) = Put\,[Z\; big] = C(\Pi(c) | c\; in\; E_3(x)). \tag{16}$$

The decision maker wishes to get an $x \geq 0$ to maximize $g_3(x)$.

Therefore, the decision maker wishes to choose an undominated solution to

$$\max_{x \geq 0}(g_1(x), g_2(x), g_3(x)). \tag{17}$$

We have translated the problem into a multiobjective game versus Nature. The first and third goals are independent but the first and second, or the second and third, goals may be conflicting.

There are various methods of exploring the Pareto optional set for the multiobjective problem in equation (17). One common procedure is to solve (Geoffrion, 1968)

$$\max_{x \geq 0}(\lambda_1 g_1(x) + \lambda_2 g_2(x) + \lambda_3 g_3(x)), \tag{18}$$

for various values of the $\lambda_i > 0, \lambda_1 + \lambda_2 + \lambda_3 = 1$. Any such solution will automatically be undominated. Another approach would be to introduce fuzzy goals for each $g_i(x)$ and take their intersection (Bellman and Zadeh, 1970). Of course, sensitivily analysis could be performed on the r_k values.

4 Summary and Conclusions

We began with considering what might be called a solution to a multiobjective programming problem whose parameters take on uncertain values. Then the first thing to discuss was how to model the uncertainty. We presented our general method of modelling uncertainty, based on t-norms and co-t-norms, which contains probability theory and possibility theory as special cases. All that was needed for this paper was a method for constructing joint uncertainty distributions and a procedure, using these joint distributions, of calculating the uncertainty of various subsets of R^m. A more detailed discussion of the algebra of uncertain variables may be found in Buckley (1990c).

We then argued that it is natural to think of the decision problem as a game versus Nature. The solution method should not depend on your model of uncertainty. We also argued that it is natural for the decision maker to consider different goals such as: (1) make the decision variables feasible and undominated; and (2) make the objective functions as large as possible. The result was that a solution to the uncertain multiobjective programming problem will be an undominated solution to a multiobjective game against Nature. An algorithm for generating these undominated solutions will depend on your model of uncertainty.

So, in solving uncertain mathematical programming problems the solution model is most important which should be independent of your model of uncertainty. Your model of uncertainty will effect your procedure for computing solutions and your interpretation of a solution. Of course, actual solutions will vary if you change your model of uncertainty (see Buckley, 1990b).

REFERENCES

Bellman, R.E. and L.A. Zadeh (1970) Decision-making in a Fuzzy Environment. Management Science 17, 141-164.

Buckley, J.J. (1989) Modelling Uncertainty in Operations Research: the Discrete Case. In J.L. Verdegay and M. Delgado (Eds.): Interfaces Between Artificial Intelligence and Operations Research, Verlag TÜV Rheinland, Koln, W. Germany, pp. 13-33.

Buckley, J.J. (1990a) Stochastic Versus Possibilistic Programming. Fuzzy Sets and Systems 34, 173-177.

Buckley, J.J. (1990b) Stochastic Versus Possibilistic Multiobjective Programming. In R. Slowinski and J. Teghem (Eds.): Stochastic Versus Fuzzy Approaches to Multiobjective Mathematical Programming Under Uncertainty, Kluwer Academic Publishers, Dordrecht (to appear).

Buckley, J.J. (1990c) A General Theory of Uncertainty Based on Co-t-Norms. Fuzzy Sets and Systems (under review).

Geoffrion, A.M. (1968) Proper Efficiency and the Theory of Vector Maximization. J. Math. Anal. Appl. 22, 618-630.

R. D. Luce and H. Raiffa (1957) Games and Decisions. Wiley, New York.

Sugeno, M. (1974) Theory of Fuzzy Integrals at its Applications. Ph.D. Thesis, Tokyo Institute of Technology, Tokyo.

L.A. Zadeh (1975) The Concept of a Linguistic VAriable and its Application to Approximate Reasoning 1, Info. Sciences 8, 199-249.

L.A. Zadeh (1978) Fuzzy Sets as a Basis for a Theory of Possibility. Fuzzy Sets and Systems 1, 3-28.

APPENDIX

In this appendix we briefly report on our procedure for extending a co-t-norm C to a mapping of S_o into $[0, +\infty)$, where S_o is a subset of $B[R^m_+]$ = all bounded, non-negative functions defined on R^m.

Let $Q = \{(x_1, \ldots, x_k, 0, 0, \ldots) | x_i \epsilon [0,1], 1 \leq i \leq k\}$. S_o and S_1 will be subsets of $B[R^m_+]$ so that given $f \epsilon S_o$ there is a sequence f_i in S_1 so that $f_i \xrightarrow{c} f$. The "c" stands for the type of convergence.

Our basic example is for $S_o \subset B[R^m_+]$ and S_1 is all simple functions in $B[R^m_+]$ with "c" denoting monotone increasing convergence.

Let $\psi : S_1 \rightarrow Q$. If $f \epsilon S_o, f_i \epsilon S_1, f_i \xrightarrow{c} f$, then let $\psi(f_i) = f^*_i = (x_{i1}, \ldots, x_{i,n_i})$ with x_{ij} in [0,1]. If C in any co-t-norm, we see that $C(f^*_i)$ is defined whereas $C(f_i)$ is not defined.

If $f \epsilon S_o$, let $M = sup|f|$. Set $\bar{f} = (1/M)f$ when $M > 1$ and $\bar{f} = f$ otherwise. We assume that $\bar{f} \epsilon S_o$ whenever $f \epsilon S_o$.

We say C is well-defined on S_o if and only if given $f \epsilon S_o, f_i \xrightarrow{c} f$ with f_i in $S_1, lim\, C(f^*_i)$ exists and is independent of the sequence f_i in S_1 which converges to f. If C is well-defined on S_o, then $C(\bar{f}) = lim\, C(f^*_i)$ and $C(f) = MC(\bar{f})$ when $M > 1$ and $C(f) = C(\bar{f})$ otherwise.

$\mathcal{A}$ will be a collection of subsets of R^m so that $f\chi_E \epsilon S_o$ whenever $f \epsilon S_o$ and $E \epsilon \mathcal{A}$. Then $C(f\chi_E)$ is defined for f in S_o and E in $\mathcal{A}$.

Example 1. How do we get possibility theory? S_o is $B[R^m_+]$ and let

$$f_i(x) = \sum_{j=1}^{n_i} a_{ij}\chi_{E_{ij}}, \tag{1}$$

so that $0 \leq f_i \uparrow \bar{f}$ with $a_{ij} \epsilon [0,1]$. Define

$$\psi(f_i) = (a_{i1}, \ldots, a_{i,n_i}), \tag{2}$$

and choose C = max. Then C is well-defined on $S_o, \mathcal{A}$ is all subsets of R^m, and

$$C(f\chi_E) = sup\,(f(x)|x \epsilon E). \tag{3}$$

Now let f be a normalized fuzzy subset of R^m and we get possibility theory.

Example 2. How do we get probability theory? S_o in all measurable f in $B[R^m_+]$ with $\int f(x)dx \leq 1$. The f_i are simple functions as in equation (1) but

$$\psi(f_i) = (a_{i1}\mu[E_{i1}], \ldots), \tag{4}$$

with μ = lebesgue measure. Choose C = LOR and $\mathcal{A}$ = measurable subsets of R^m. Then C is well-defined on S_o and

$$C(f\chi_E) = \int_E f(x)dx. \tag{5}$$

If $f \epsilon S_o$ with its integral equal to one, then we obtain probability theory.

APPROACHING FUZZY INTEGER LINEAR PROGRAMMING PROBLEMS

F. Herrera and J.L. Verdegay
Departamento de Ciencias de la Computacion
e Inteligencia Artificial
Universidad de Granada
18071 Granada (Spain)

Abstract: Integer Programming problems in which they exists some lack of precision of a vague nature in their formulation are considered. It is also provided a classification of this kind of problems, and a description for each of them is carried out. In particular, we focuse on the fuzzy boolean case for which a solution method is shown.

Keywords: Fuzzy integer programming; fuzzy constraint; fuzzy numbers; Fuzzy Boolean programming.

1. INTRODUCTION

Linear Programming (LP) problems with discrete (integer and/or 0-1) variables are of utmost importance in many fields as, e.g., those related to artificial intelligence, operations research and management. They are specially important for representing and reasoning with propositional knowledge, but, unfortunately, not very much research work has been done on their fuzzification ([3], [16]).

The aim of this paper is to approach the fuzzification of this kind of problems to provide a tool for helping reason in imprecise-knowledge-based systems by focusing, particularly, on boolean fuzzy LP (BFLP) problems.

As it is well known, from a conventional LP problem

$$\begin{aligned} &\text{Max: } z = cx \\ &\text{s.t:} \\ &\qquad Ax \le b \\ &\qquad x \ge 0 \end{aligned} \tag{1}$$

one can obtain different canonical fuzzy LP (FLP) problems according to whether the fuzzification affect the objective, the constraints, or the coefficients defining the constraint set.

Roughly speaking, an integer LP (ILP) problem is a LP problem in which the variables are restricted to take on integer values. Hence, concert with the above fuzzification corresponding to theLP problems, now related to ILP problems, it has a new one: to fuzzify the integer nature of the variables, that is, to permit variables with almost integer values, [3].

Traditionally, conventional ILP problems may be divided into two great groups. First, the one corresponding to the 0-1 LP problems, that is the LP problems in which the variables can only take on the values 0 (false) or 1 (true). Usually, these are called boolean LP problems. On the other hand, those formerly related to ILP problems, that is problems in which some of the variables are restricted to take on integer, non necessarily 0 or 1, values. These are the so called integer mixed LP problems.

From this point of view, and taking into account the above four possible canonical ways to fuzzify the ILP problem, one can also consider two different classes of fuzzy ILP (FILP) problems. First, the one corresponding to some fuzzification of Boolean LP problems, and the one derived from the fuzzification of some element characterizing an integer mixed LP problem.

As said before, here we focus on the BFLP problems in which the variables take on the values 0 and 1, but the constraints are imprecisely defined. The next section presents a classification for the FILP problems. Then, the BFLP problems are considered in section 3, in which an algorithm to solve this kind of problems is provided. Finally, a numerical example is shown.

2. CLASSIFYING FUZZY INTEGER LINEAR PROGRAMMING PROBLEMS

The overall situation corresponding to general FILP problems, not the 0-1 problems, can be described by the following example.

Ilustrative example: In an Andalusian hospital, four different types of special analyses are performed daily.A study carried out about the daily demand of analyses is shown in the following table.

Type of Analysis	Amount
1	450
2	900
3	360
4	300

Daily these results have some variations that at most reach around the 20%, 23%, 25% and 20% of each of the respective amount of analyses carried out.

The staff of the hospital plans to automate the Analysis Service, and has two possible equipment of similar quality to acquire, whose costs will finally depend on technical improvements.

Equipment A			Equipment B	
Cost	around 3.5 millions		around 3 millions	
Performance			Performance	
Type of Analysis		Amount/day	Type of Analysis	Amount/day
	1	42	1	45
	2	72	2	39
	3	39	3	18
	4	27	4	36

It is required to determine the number of equipments of type A and B, to buy as for satisfying the demand by minimizing the inversion.

This is a shortened version of a more large problem to be solved in the Andalusian School of Public Health. It is clear that both the constraints and the coefficients in the objective are imprecisely given, and even the daily amount of analyses of the different types can be seen in this way. Therefore, to solve this problem some model of FILP problem is needed.

As it has been said before, for the FILP problems two main groups can be considered. The first one is that constituted by the problems in which the variables are restricted to take on the integer values (different from 0 and 1), and the second one is that involving the boolean values for the variables. Next, each of them will be describe.

a.1) FILP problems with imprecisely defined constraints. Similarly to the conventional case, these problems can be written as

$$\begin{aligned}
&\text{Max: } z = cx\\
&\text{s.t:}\\
&\qquad \sum_{j\in N} a_{ij}x_j \lesssim b_i, \quad i \in M = \{1,\dots,m\} \qquad (2)\\
&\qquad\qquad x_j \geq 0, \quad j \in N = \{1,\dots,n\}\\
&\qquad\qquad x_j \in \mathbb{N}, \quad j \in I \subseteq N
\end{aligned}$$

where $\mathbb{N}$ is the set of integer numbers, $c\in\mathbb{R}^n$ and a_{ij}, $b_i\in\mathbb{R}$, $i\in M$, $j\in N$. The symbol $\lesssim$ means that the decision-maker is willing to permit some violations in the accomplishment of the constraints, that is, he considers fuzzy constraints defined by membership functions

$$\mu_i: \mathbb{R}^n \longrightarrow (0,1], \; i \in M \qquad (3)$$

each of them give the degree to each $x\in\mathbb{R}^n$ accomplishes the respective constraint.

Now, by considering the set of indices I, two different problems arise. If I = N, one has the Fuzzy Pure Integer LP problem. If $I \subset N$, then (2) is a Fuzzy Mixed Integer LP problem.

a.2) FILP problems with imprecise coefficients in the objective function, that is, with coefficients defined by fuzzy numbers. The problem can be written as

$$\begin{aligned}
&\text{Max: } z = \sum_{j\in N} c_j x_j\\
&\text{s.t:}\\
&\qquad \sum_{j\in N} a_{ij}x_j \leq b_i, \quad i \in M \qquad (4)\\
&\qquad\qquad x_j \geq 0, \quad j \in N\\
&\qquad\qquad x_j \in \mathbb{N}, \quad j \in I \subseteq N
\end{aligned}$$

where the novelty, with respect to (2), is that all the c_j are fuzzy numbers and the constraint set is a conventional set. Thus one has the membership functions

$$\mu_j: \mathbb{R} \longrightarrow (0,1], \; j \in N \qquad (5)$$

expressing the lack of precision on the values of the coefficients that the decision-maker has.

By considering the set of indices I, two different problems arise again. If I = N, the Fuzzy Pure Integer LP problem is obtained. If $I \subset N$, then (4) is a Fuzzy Mixed Integer LP problem. Let it note that formally these two FILP problems are hardly different from the corresponding ones in a.1.

a.3) FILP problems with fuzzy numbers as coefficients of the technological matrix and the right-hand-side, that is, one has

$$\begin{aligned} &\text{Max: } z = cx \\ &\text{s.t:} \\ &\quad \sum_{j\in N} a_{ij}x_j \lesssim b_i,\ i \in M \\ &\qquad x_j \geq 0,\ j \in N \\ &\qquad x_j \in \mathbb{N},\ j \in I \subseteq N \end{aligned} \tag{6}$$

where a_{ij}, $b_i \in T(\mathbb{R})$, the set of triangular fuzzy numbers, and $c \in \mathbb{R}^n$. The symbol $\lesssim$ means, like in (2), that the decision-maker permits some flexible accomplishment for the constraints. Thus, the following membership functions are to be considered (see Fedrizzi et al. in this book),

$$\forall b_i,\ i\in M,\ \mu_i : \mathbb{R} \longrightarrow (0,1] \tag{7}$$

$$\forall a_{ij},\ (i,j)\in M\times N,\ \mu_{ij} : \mathbb{R} \longrightarrow (0,1] \tag{8}$$

$$\forall i\in M,\ \mu^i : T(\mathbb{R}) \longrightarrow (0,1],\ i \in M \tag{9}$$

where the μ^i are membership functions like the one in(3).

Then, according to either $I = N$ or $I \subset N$, for (6) one can obtain Fuzzy Pure Integer LP problems and Fuzzy Mixed Integer LP problems respectively.

a.4) FILP problems in which the variables can take values almost integer [3], that is, this problem,

$$\begin{aligned} &\text{Max: } z = cx \\ &\text{s.t:} \\ &\quad \sum_{j\in N} a_{ij}x_j \leq b_i,\ i \in M \\ &\qquad x_j \geq 0,\ j \in N \\ &\qquad x_j \in F(\mathbb{N}),\ j \in I \subseteq N \end{aligned} \tag{10}$$

where all the elements taking part into the problem are conventional ones, but now $x_j \in F(\mathbb{N})$ is to be meant that the decision-maker is willing to consider some violation in the accomplishment of the property of variables of being integer. Examples of this kind of problems can be found in [3].

Another great group of FILP problems is the one derived from the case in which the variables only take on the values 0 or 1, that is, the BFLP problems. Since the meaning of the fuzzy elements that can take part into these last problems is the same than above, only the corresponding models will be written. On the other hand, now is not necessary to distinguish the pure case from the mixed one because, as it is well known, any integer variable can be expressed as a combination of the boolean ones. Thus one has:

b.1) BFLP problems with imprecise defined constraints:

$$\begin{aligned} &\text{Max: } z = cx \\ &\text{s.t:} \\ &\quad \sum_{j\in N} a_{ij}x_j \lesssim b_i,\ i \in M \\ &\quad\quad x_j \in \{0, 1\},\ j \in N \end{aligned} \tag{11}$$

b.2) BFLP problems with coefficients in the objective function defined by fuzzy numbers:

$$\begin{aligned} &\text{Max: } z = \sum_{j\in N} c_j x_j \\ &\text{s.t:} \\ &\quad \sum_{j\in N} a_{ij}x_j \le b_i,\ i \in M \\ &\quad\quad x_j \in \{0, 1\},\ j \in N \end{aligned} \tag{12}$$

and finally,

b.3) BFLP problems with fuzzy numbers representing the coefficients of both the technological matrix and the right-hand-sides:

$$\begin{aligned} &\text{Max: } z = cx \\ &\text{s.t:} \\ &\quad \sum_{j\in N} a_{ij}x_j \lesssim b_i,\ i \in M \\ &\quad\quad x_j \in \{0, 1\},\ j \in N \end{aligned} \tag{13}$$

Of course, for the FILP problems as well as the BFLP ones, combinations between the presented models can be allowed.

Remark 1. It is very important to quote that BFLP problems with almost integer variables are nothing but conventional LP problems with bounded variables. Therefore, they do not constitute any special case of BFLP problems.

Most applications of the BFLP problems can be considered in knowledge-based systems in which is necessary to represent and reason with propositional knowledge [17]. From this point of view and because of the relevance of its applications, the next section focuses on the case b.1, for which a solution method is shown.

3. SOLVING BOOLEAN FUZZY LINEAR PROGRAMMING PROBLEMS

Considering a general 0-1 fuzzy mathematical programming problem of type b.1

$$\begin{aligned} &\text{Min: } z = f(x) \\ &\text{s.t:} \\ &\qquad g_i(x) \lesssim b_i,\ i \in M \\ &\qquad x_j \in \{0,1\},\ j \in N \end{aligned} \qquad (14)$$

For each constraint, denote

$$X_i = \{x \in \mathbb{R}^n /\ g_i(x) \lesssim b_i,\ x_j \in \{0,1\}\}$$

Then, the constraint set can be represented by

$$X = \bigcap_{i=1}^{m} X_i$$

and (14) can be rewritten as

$$\text{Min } \{z = f(x) /\ x \in X\} \qquad (15)$$

As it is known, the fuzziness in (14) implies that for each constraint one has a membership function $\mu_i : \mathbb{R}^n \longrightarrow (0,1]$ being continuous and monotonic, $i \in M$, and giving the accomplishment degree to which $x \in \mathbb{R}^n$ satisfies the i-th fuzzy constraint,

$$\mu_i(x) = \begin{cases} 1 & \text{if} \quad g_i(x) \le b_i \\ f_i(g_i(x)) & \text{if} \quad b_i \le g_i(x) \le b_i + d_i \\ 0 & \text{if} \quad g_i(x) \ge b + d_i \end{cases}$$

where the d_i, $i \in M$, is the margin with which the decision-maker permits that the i-th constraint to be violated.

Thus, it is clear that an α-cut, $\forall \alpha \in (0,1]$, of the constraint set may be written as

$$X(\alpha) = \{x \in \mathbb{R}^n \ /\ \mu_X(x) \ge \alpha\}$$

where $\forall x \in \mathbb{R}^n$, $\mu_X(x) = \inf \{\mu_i(x), i \in M\}$. Also, $X_i(\alpha)$ will denote an α-cut of the i-th constraint, $i \in M$

Then, as $\forall \alpha \in (0,1]$

$$X(\alpha) = \bigcap_{i=1}^{m} X_i(\alpha) = \bigcap_{i=1}^{m} \{x \in \mathbb{R}^n / \mu_i(x) \geq \alpha\} =$$

$$= \bigcap_{i=1}^{m} \{x \in \mathbb{R}^n / f_i[g_i(x)] \geq \alpha, x_j \in \{0,1\}\} =$$

$$= \bigcap_{i=1}^{m} \{x \in \mathbb{R}^n / g_i(x) \leq f_i^{-1}(\alpha), x_j \in \{0,1\}\} =$$

$$= \bigcap_{i=1}^{m} \{x \in \mathbb{R}^n / g_i(x) \leq r_i(\alpha), x_j \in \{0,1\}\}$$

with $r_i(\alpha) = f_i^{-1}(\alpha)$, (14) can be expressed as

$$\begin{aligned} &\text{Min: } z = f(x) \\ &\text{s.t:} \\ &\qquad g_i(x) \leq r_i(\alpha),\ i \in M \\ &\qquad x_j \in \{0,1\},\ \alpha \in (0,1] \end{aligned} \tag{16}$$

Now, denoting, $\forall \alpha \in (0,1]$, $S(\alpha) = \{x \in \mathbb{R}^n / z(\alpha) = \text{Min } f(x),\ x \in X(\alpha)\}$, a fuzzy solution to (14) can be defined, [10],

Definition: A fuzzy solution to (14) is the fuzzy set with the membership function

$$\lambda(x) = \begin{cases} \sup\limits_{x \in S(\alpha)} \alpha & x \in \bigcup\limits_{\alpha} S(\alpha) \\ 0 & \text{elsewhere} \end{cases}$$

Consider, in particular, the following BFLP problem in which without, loss of generality, one supposes thatall the coefficients are integer numbers.

$$\begin{aligned} &\text{Min: } z = cx \\ &\text{s.t:} \\ &\qquad a_i x \lesssim b_i,\ i \in M \\ &\qquad x_j \in \{0,1\},\ j \in N \end{aligned} \tag{17}$$

The membership functions of the constraints are defined by

$$\mu_i(x) = \begin{cases} 1 & \text{if} \quad a_i x \le b_i \\ [(b_i+d_i)-a_i x]/d_i & \text{if} \quad b_i \le a_i x \le b_i+d_i \\ 0 & \text{if} \quad a_i x \ge b_i+d_i \end{cases}$$

Hence, (17) can be written as

$$\begin{aligned} &\text{Min: } z = cx \\ &\text{s.t:} \\ &\qquad a_i x \le b_i + d_i(1-\alpha),\ i \in M \qquad (18) \\ &\qquad x_j \in \{0,1\},\ \alpha \in (0,1],\ j \in N \end{aligned}$$

which is a conventional parametric 0-1 LP problem, in the following, for each fixed $\alpha \in (0,1]$, will be represented by $P(\alpha)$ and, correspondingly, $x(\alpha)$ will denote its optimal solution.

To approach a solution method for (17) which provides a fuzzy solution, in the following, $P(\alpha)$ will be used as an intermediate, and $[x]$ will denote the largest integer less than or equal to x. Thus, if $\{x\}$ denotes the fractional remainder of $x \in \mathbb{R}$, then $x = [x]+\{x\}$, $0 \le \{x\} < 1$.

Proposition 1. Let $x(\alpha')$ an the optimal 0-1 solution for $P(\alpha')$ for some fixed $\alpha' \in (0,1]$. Then

$$\lambda(x(\alpha')) = \min \{\mu\ (x(\alpha')),\ i \in M\}$$

Proof. Let's denote

$$\theta = \min \{\mu_i(x(\alpha')),\ i \in M\}$$

Then, $\theta \ge \alpha'$ and $X(\theta) \subseteq X(\alpha')$. As

$$cx(\alpha') = \text{Min } \{cx/\ x \in X(\alpha')\}$$

and $x(\alpha') \in X(\theta) \subseteq X(\alpha')$, it followes that

$$cx(\alpha') = \text{Min } \{cx/\ x \in X(\theta)\}$$

Moreover, as $\forall\ \alpha > \theta$, one has $x(\alpha') \notin X(\alpha)$, and then

$$\lambda(x_{\alpha'}) = \theta = \min \{\mu_i(x(\alpha')),\ i \in M\} \qquad \blacksquare$$

Corollary. Let $x(\alpha')$ be an optimal 0-1 solution for $P(\alpha')$ for some fixed $\alpha' \in (0,1]$. Let $\theta = \lambda(x(\alpha'))$ be. Then $x(\alpha')$ is an optimal solution for $P(\alpha)$, $\forall \alpha \in [\alpha',\theta]$.

Proof. $\forall \alpha \in [\alpha',\theta]$, one has $X(\theta) \subseteq X(\alpha) \subseteq X(\alpha')$. Therefore, as $x(\alpha') \in X(\theta)$ and

$$cx(\alpha') = \text{Min } \{cx/\ x \in X(\alpha')\}$$

then $cx(\alpha')= \text{Min}\ \{cx/\ x\in X(\alpha)\}$. ∎

Remark 2. Notice from the above result that

$$\forall\alpha\in[\alpha',\theta],\ x(\alpha') = x(\alpha)$$

and, in particular if $\alpha' \neq \alpha$, then $\lambda(x(\alpha')) = \lambda(x(\alpha)) = \theta$.

Remark 3. We will need the following relation. Let's consider, $\forall\alpha\in(0,1]$,

$$1_\alpha = \begin{cases} 1 & \text{if } \max\limits_i \{b_i+d_i(1-\alpha)\}=0 \\ \min\limits_i \left\{\{b_i+d_i(1-\alpha)\}:\ \{b_i+d_i(1-\alpha)\}\neq 0\right\} & \text{elsewhere} \end{cases}$$

Let $d' = \max\ \{d_i\}$, and denote $\Delta\alpha = 1_\alpha/d'$. Then, it is evident that

$$d_i\cdot\Delta\alpha \leq 1,\ \forall i\in M.$$

On the other hand, $\forall i\in M$ such that $\{b_i+d_i(1-\alpha)\}\neq 0$, one has

$$1 > \{b_i+d_i(1-\alpha)\} - d_i\cdot\Delta\alpha = \{b_i+d_i(1-\alpha)\}- d_i\cdot 1_\alpha/d' \geq$$

$$\geq \{b_i+d_i(1-\alpha)\} - d_i\cdot 1_\alpha/d_i = \{b_i+d_i(1-\alpha)\} - 1_\alpha \geq 0$$

and therefore

$$[b_i+d_i(1-\alpha)] = [b_i+d_i(1-\alpha) - d_i\cdot\Delta\alpha] = [b_i+d_i(1 - (\alpha+\Delta\alpha))]$$

Proposition 2. Let $\theta = \lambda(x(\theta))\in(0,\ 1)$, and $\alpha' = \theta+\Delta\theta$, with $\Delta\theta$ as previously defined. Consider $\alpha'' = \lambda(x(\alpha'))$. Then, $\forall\alpha\in(\theta,\ \alpha'']$, $x(\alpha')$ is an optimal solution for $P(\alpha)$.

Proof. Let's consider $(\theta,\alpha'']$ divided into two subintervals: (θ,α') and $[\alpha',\ \alpha'']$. The proposition will be shown for each of them. In the second subinterval the proof is immediate because of the above corollary.

In the other subinterval, let $\beta\in(\theta,\ \alpha')$ and $x(\beta)$ be an optimal 0-1 solution for $P(\beta)$. If $\beta' = \lambda(x(\beta))$, then $\beta' \geq \beta > \theta$.

As $\beta' = \lambda(x(\beta))$, $\exists i\in M$ such that one verifies

$$a_i x(\beta) = b_i + d_i(1-\beta') < b_i + d_i(1-\theta)$$

being $a_i x(\beta)$ an integer number.

Moreover, from the integer nature of the coefficients,

$[b_i + d_i(1-\beta')] = b_i + d_i(1-\beta')$

Then there are two possibilities:

a) $[b_i + d_i(1-\theta)] = b_i + d_i(1-\theta)$. But as

$b_i + d_i(1-\beta') < b_i + d_i(1-\theta)$ (integer values)

and $d_i \cdot \Delta\theta \le 1$, then

$$b_i + d_i(1-\beta') \le b_i + d_i(1-\theta) - d_i \cdot \Delta\theta = b_i + d_i(1 - (\theta+\Delta\theta)) =$$

$$= b_i + d_i(1-\alpha') \Rightarrow \beta' \ge \alpha'$$

b) $[b_i+d_i(1-\theta)] \ne b_i+d_i(1-\theta)$. But as

$d_i \cdot \Delta\theta \le \{b_i+d_i(1-\theta)\}$

then

$$b_i + d_i(1-\beta') \le [b_i + d_i(1-\theta)] + \{b_i + d_i(1-\theta)\} - d_i \cdot \Delta\theta =$$

$$= b_i+d_i(1-\alpha') \Rightarrow \beta' \ge \alpha'$$

Since $\beta' \ge \alpha'$ and $\alpha' > \beta$, then $\alpha' \in [\beta,\beta']$. Thus, from the corollary, $x(\beta)$ is an optimal solution for $P(\omega)$, $\forall\omega \in [\beta,\beta']$. Hence, $x(\beta)$ is also an optimal solution for $P(\alpha')$, thus, $cx(\beta) = cx(\alpha')$, it results that $x(\alpha')$ is an optimal solution for $P(\omega)$, $\forall\omega \in [\beta, \beta']$.

Therefore, $x(\alpha')$ is an optimal solution for $P(\beta)$, $\forall\beta \in (\theta,\alpha')$, and so, finally, $x(\alpha')$ is an optimal solution for $P(\alpha)$, $\forall\alpha \in (\theta,\alpha'']$. ■

The above results serve to built an algorithm to solve the former problem (17) by using (18) as an auxiliar problem. The algorithm works as follow: First $P(0)$, $\alpha_1 = 0$, is solved by using the Glover's Enumeration Scheme [17], and then one determines the maximum value of α, called θ, for which the solution remains optimal in the interval $[\alpha_1, \theta]$. If $\theta \ne 1$, then we consider $\alpha = \theta+\Delta\theta$ and solve $P(\alpha)$ looking for an interval to α again. The process is repeated until $\theta = 1$ is reached.

The interval of values α for which x is an optimal solution for $P(\alpha)$ will be denoted $I_{x(\alpha)}$. It is clear that $I_{x(\alpha)} \subseteq [0,1]$. The algorithm is the following.

Algorithm

Suppose a BFLP problem like (17). Then,

Step 0: Let $\alpha = \alpha_1 = 0$.
Step 1: Solve $P(\alpha)$. Let $x(\alpha)$ be an optimal solution for $P(\alpha)$.

Step 2: Let $\theta = \lambda(x(\alpha)) = \min \{\mu_i(x_\alpha),\ i \in M\}$.
If $\alpha_1 = 0$
Then $x(\alpha)$ is an optimal solution for $P(\alpha)$ $\forall \alpha \in [\alpha_1, \theta]$
Otherwise $x(\alpha)$ is an optimal solution for $P(\alpha)$ $\forall \alpha \in (\alpha, \theta]$

Step 3: If $\theta < 1$
Then $\alpha_1 = \theta$, $\alpha = \theta + \Delta\theta$, go to Step 1

Step 4: Stop.

The following example illustrates the algorithm. Let's consider

$$\begin{aligned} \text{Min: } & z = -10x_1 - 7x_2 - 15x_3 \\ \text{s.t: } & 3x_1 + 2x_2 + x_3 \lesssim 4 \\ & 10x_2 + 5x_3 \lesssim 15 \\ & 2x_1 + x_3 \lesssim 2 \\ & x_1, x_2, x_3 \in \{0,1\} \end{aligned}$$

Let $d_1 = 3$, $d_2 = 3$ and $d_3 = 2$, the margins that the decision-maker tolerates in the accomplishment of each constraint. Then,

Step 0: $\alpha = \alpha_1 = 0$.
Step 1: $x(0) = (1, 1, 1)$ optimal solution for $P(0)$.
Step 2: $\theta = .333$, $I_{x(0)} = [0, .333]$.
Step 3: $\theta < 1$
Then $\alpha_1 = .333$, $\alpha = \theta + \Delta\theta = .444$, go to Step 1.
Step 1: $x(.444) = (1, 0, 1)$, optimal solution for $P(.444)$.
Step 2: $\theta = 0.5$, $I_{x(.444)} = (.333, 0.5]$.
Step 3: $\theta < 1$
Then $\alpha_1 = 0.5$, $\alpha = \theta + \Delta\theta = .666$, go to Step 1.
Step 1: $x(.666) = (0, 1, 1)$ optimal solution for $P(.666)$.
Step 2: $\theta = 1$, $I_{x(.666)} = (0.5, 1]$.
Step 3: $\theta = 1$
Step 4: Stop.

Thus one has,

$x(0) = (1, 1, 1)$, $I_{x(0)} = [0, .333]$

$x(.444) = (1, 0, 1)$, $I_{x(.444)} = (.333, 0.5]$

$x(.666) = (0, 1, 1)$, $I_{x(.666)} = (0.5, 1]$

and finally, the fuzzy solution obtained finally is the fuzzy set

$$\tilde{S} = \{ (1,1,1)/0.333,\ (1,0,1)/0.5,\ (0,1,1)/1 \}$$

4. CONCLUSIONS

The FILP and BFLP models provide an useful framework to solve a lot of problems from different fields. In this paper the BFLP problem has been focused, as a first step, on the development of Mathematical Programming models to represent and reason in fuzzy-logic-based decision support systems.

There is, however, a hard work to do. For instance, on the one hand, with respect to the FILP problems, methods to solve them are needed, and then possible relations between their fuzzy solutions are to be considered. On the other hand, for the BFLP problems, also is to be reserched the relation between the fuzzy solution here obtained and the provided from other approaches.

Acknowledgement

We are indebted to Ms. Guadalupe Carmona from the Andalusian School of Public Health for providing us the data for the example in Section 2.

References

[1] Bellman, R.E.. and Zadeh, L.A. (1970): Decision Making in a Fuzzy Environment. Man. Sci. 17 (B), 4, 141-164.

[2] Delgado, M., J.L. Verdegay and M.A. Vila (1989): A General Model for Fuzzy Linear Programming. Fuzzy Sets and Systems 29, 21-29.

[3] Fabian, C. and M. Stoica (1984): Fuzzy Integer Programming. In Fuzzy Sets and Decision Analysis. H.J. Zimmermann, L. A. Zadeh and B. R. Gaines (Eds). North-Holland, 123-131.

[4] Gal, T. (1979): Postoptimal Analyses, Parametric Programming and Related Topics. McGraw Hill.

[5] Garfinkel, R.S and Nemhauser, G.L. (1972): Integer Programming. Wiley Interscience.

[6] Geoffrion, A. M. and R. Nauss (1977): Parametric and Postoptimality Analysis in Integer Linear Programming". Management Science 23, 453-466.

[7] Marsten, R.E. and Morin, T.L. (1977): Parametric Integer Programming: The Right-Hand-Side Case. Annals of Discrete Math. 1, 375-390.

[8] Nauss, R. M. (1979): Parametric Integer Programming. University of Missouri Press.

[9] Negoita, C.V. and D.A. Ralescu (1975): Applications of Fuzzy Sets to Systems Analysis. Birkhauser.

[10] Orlovski, S.A. (1977): On Programming with Fuzzy Constraint Sets. Kybernetes 6, 197-201.

[11] Taha, H.A. (1975): Integer Programming. Theory, Applications and Computations. Academic Press.

[12] Verdegay, J.L. (1982): Fuzzy Mathematical Programming. In Fuzzy Information and Decision Processes. M.M. Gupta and E. Sanchez (Eds). North-Holland, 231-237.

[13] Yager, R.R. (1988): A Mathematical Programming Approach to Inference with the Capability of Implementing Default Rules. Int. J. of Man-Machine Studies 29, 685-714.

[14] Zadeh, L.A. (1978): Fuzzy Sets as a Basis for a Theory of Possibility. Fuzzy Sets and Systems, 1, 3-28.

[15] Zadeh, L. A. (1979): Calculus of Fuzzy Restrictions. In Fuzzy Sets and their Applications to Cognitive and Decision Processes. L.A. Zadeh, K.S. Fu, K. Tanaka and M. Shimura, (Eds.), Academic Pres, 1-40.

[16] Zimmermann, H.J. and M.A. Pollatschek (1984): Fuzzy 0-1 Linear Programs. In Fuzzy Sets and Decision Analysis. H. J. Zimmermann, L.A. Zadeh and B. R. Gaines (Eds). North-Holland, 133-145.

[17] Zimmermann, H.J. (1987): Fuzzy Sets, Decision Making and Expert Systems. Kluwer Academic Publishers.

Interactive Bicriteria Integer Programming: A Performance Analysis

R. Ramesh†
M. H. Karwan‡
S. Zionts†

†School of Management,
State University of New York at Buffalo,
Buffalo, N.Y. 14260.

‡Department of Industrial Engineering,
State University of New York at Buffalo,
Buffalo, N.Y. 14260.

Abstract

The bicriteria problem is an important problem in multiobjective optimization that has been extensively studied in the literature. Practical applications and algorithmic investigations of this problem are numerous. We have developed an efficient interactive solution framework for solving bicriteria integer programming problems. In this framework, a decision maker is assumed to have an implicit utility function of the objectives he wishes to maximize, and the function is considered to be pseudoconcave and nondecreasing. The decision-maker's preference structure is assessed using pairwise comparison questions, and the problem is solved by an interactive branch-and-bound method, in which the decision maker participates in the solution process. The bicriteria problem has several special properties, and the solution framework employs several algorithmic components that exploit these properties leading to an efficient solution process. In this article, we report on a detailed performance investigation carried out on the solution framework. The results show that the proposed approach is viable for solving practical problems of medium size with nonlinear utility functions.

Keywords:
Bicriteria Optimization, Branch-and-Bound, Integer Programming, Implicit Utility Functions, Interactive Methods.

1 Introduction

This paper presents a performance analysis of an interactive solution framework for bicriteria integer programming. A decision maker is assumed to have only an implicit utility function of the two objectives he wishes to maximize. The problem is solved by an interactive branch-and-bound method in which the decision-maker's preference structure is explored and determined using pairwise comparison questions. The proposed framework exploits the structure and special properties of the bicriteria problem. A number of new concepts on bicriteria integer programming that offer great efficiency in the solution process are used in this framework. In this article, we report on a detailed investigation on the performance characteristics of this framework carried out by us.

Literature on the development of theory and algorithms for multicriteria mathematical programming is extensive. In particular, these problems have been addressed using three approaches: vector optimization, maximization of an explicit utility function derived using the multiattribute utility theory, and the maximization of an implicit utility function using interactive methods. A comprehensive survey of the multicriteria mathematical programming methods is provided in Evans [1984]. In particular, Teghem and Kunsch [1986] survey the interactive methods for multicriteria integer programming. The methods of Gabbani and Magazine [1986], Marcotte and Soland [1986] and Ramesh *et al.* [1986][1989][1990] belong to this category.

A theoretical characterization of bicriteria mathematical programs is given in Geoffrion [1967]. Pasternak and Passy [1982], Benson [1979] and Chalmet *et al.* [1986] consider the bicriteria problem using the vector optimization approach. These methods obtain all the efficient solutions to a bicriteria integer programming problem. The method proposed in Ramesh *et al.* [1990] follows the implicit utility maximization approach, and obtains the most preferred solution to the problem through an interaction with the decision maker. In this regard, this method is an extension of the literature on the generation of all the efficient solutions to obtaining a workable, most preferred solution to the bicriteria problem. In this article, we examine some of the critical performance characteristics of the algorithm developed in Ramesh *et al.* [1990].

The problem of concern is formulated as follows.

$$\begin{aligned} \textbf{BCIP}: \quad & \max \quad g(Cx) \\ & \text{s.t.} \quad Ax \leq b \\ & \qquad\;\; x \geq 0, \text{ integer} \end{aligned}$$

where C and A are $2 \times n$ and $m \times n$ matrices of criteria and constraint coefficients respectively, $b \in R^m$ is the vector of resources available, $x \in R^n$ denotes the vector of decision variables and $g(\cdot)$ is the implicit utility function of the decision maker which is assumed to be pseudoconcave and nondecreasing. The linear relaxation of problem (BCIP) is denoted as (BCLP).

The organization of this paper is as follows. Section 2 briefly summarizes the methodology and the algorithmic components of the method proposed in Ramesh *et al.* [1990]. Section 3 presents the critical performance characteristics of the algorithmic components obtained from a detailed investigation, and Section 4 presents the conclusions.

2 Methodology

The problem (BCIP) is solved by employing an interactive bicriteria linear programming algorithm within a branch-and-bound framework. The bicriteria linear programming algorithm takes into account the special structure of the bicriteria problem, and is based on the method of Zionts and Wallenius [1983]. The initial step consists of relaxing the integrality constraints and solving problem (BCLP) using the bicriteria linear programming algorithm. If the optimal solution to problem (BCLP) is integer, then it is also optimal to problem (BCIP). Otherwise, a branch-and-bound search for the integer optimum is conducted. This phase is similar to the branch-and-bound search used in solving single objective integer programming problems. However, several special algorithmic components that are unique to the bicriteria problem are employed in solving the problem. We summarize these components in the following discussion.

Cuts in the Objective Function Space

The decision-maker's responses yield certain convex cones (see Ramesh *et al.* [1989]) in the objective function space that identify inferior solutions. In the bicriteria problem, it is possible to eliminate solutions dominated by some of these cones by deriving cuts on the objective functions. These cuts are simple bounds on the objectives and are derived while solving the bicriteria linear relaxations of the candidate problems.

Partitioning the Solution Space

Initial computational experience with problem (BCIP) for zero-one variables revealed that the set of feasible solutions to a candidate problem obtained after fixing many variables is almost entirely dominated by the incumbent solution, especially in cases where good initial incumbents are available. This suggested the strategy of partitioning the solution space into sets that are efficient with respect to the incumbent and search for the optimum separately in each set. In the bicriteria case, at most two such sets can be constructed, and hence the exclusive search is computationally viable. Two strategies for partitioning the solution space have been developed. Computational experience has shown that about 25% of the total number of questions asked of the decision maker can be avoided by partitioning.

Preference Structure Representation

The decision-maker's pairwise comparisons yield constraints on the weights on the objectives in the weight space and convex cones in the objective function space. The constraints on the weights are termed the λ-costraints. The λ-constraints are used to restrict the space of feasible weights that are consistent with the decision-maker's responses in generating efficient solutions. If the underlying utility function is nonlinear, then often inconsistency develops among these weights. In such cases, some of the constraints are dropped to obtain a feasible set of weights. Therefore, this results in a loss of information on the decision-maker's preference structure obtained from earlier questions. This loss is avoided by using a parallel representation of the preference structure in the objective function space using the convex cones. Accordingly, the constraints on the weights and the convex cones are used in tandem in identifying solutions that should be eliminated from consideration in the search for the most preferred solution.

Fathoming Tests

Two fathoming tests are used in the solution procedure. The first test is used to determine whether a candidate problem obtained from branching can be fathomed without exploring the decision-maker's preference structure further. The second test is employed after solving the bicriteria linear relaxation of a candidate problem to determine whether the candidate problem can be fathomed based on the decision-maker's previous responses.

Search Strategies

Ramesh *et al.* [1986] present two branch-and-bound search strategies for solving a multicriteria integer programming problem when the underlying utility function is linear. Preliminary computational experience showed that these algorithms may result in having to ask several questions of the decision maker when the underlying utility function is nonlinear. However, in solving problem (BCIP), a combination of the two strategies resulted in a significant reduction in the number of questions as well as the solution time.

BCLP Algorithm

The bicriteria problem has several special properties. The objective function space is two dimensional. Therefore, an efficient extreme point solution has at most two distinct adjacent efficient extreme point solutions in the objective function space. Furthermore, the weight space for the bicriteria problem is unidimensional. Inconsistency among the λ-constraints does not arise while solving a bicriteria problem except when a decision maker likes an efficient tradeoff from a locally optimal efficient extreme point solution to the linear relaxation (see Ramesh *et al.* [1989]). These properties have been exploited to simplify the Zionts and Wallenius procedure for the bicriteria case. Further, the procedure has been expanded to include deriving and enforcing the cuts on the objectives as well as the preference structure estimations using convex cones within the framework of bicriteria integer programming.

Search Termination Strategies

The Zionts and Wallenius procedure yields an extreme point solution that is preferred to all its adjacent extreme points. This solution is only locally optimal, because the decision maker may prefer a solution along an edge leading to an adjacent extreme point solution. This can happen when the underlying utility function is nonlinear. Therefore, the solution obtained from the Zionts and Wallenius procedure may not be directly used in a rigorous application of the fathoming criteria, unless the solution is also globally optimal. Accordingly, two heuristic procedures and a rigorous search termination strategy have been developed. The heuristics performed very well in obtaining the integer optima in the class of problems tested.

We present results from a detailed experimental investigation on the performance of the above algorithmic components for bicriteria integer programming in the following section.

3 Performance Analysis

The solution framework for problem (BCIP) has been programmed in FORTRAN V and implemented on IBM 3081/GX at the State University of New York at Buffalo. The solution framework has been extensively tested with randomly generated problems. The

Table I: Performance Statistics: Design 1 †

Problem size ($m \times n$)	Algorithm	# Questions for (BCIP)	$\frac{Q(BCIP)}{Q(BCLP)}$	Solution Time for (BCIP)	$\frac{T(BCIP)}{T(BCLP)}$
	SA1	26.85	14.79	10.09	72.32
(20 × 40)	SA2	22.05	10.12	5.67	39.48
	SA3	16.74	7.64	4.78	37.12
	SA1	31.33	18.43	19.26	93.49
(30 × 70)	SA2	27.90	15.88	15.65	61.34
	SA3	17.85	8.90	11.34	36.43

† Each cell is the average of twenty problems.

decision0maker's responses are simulated on the computer using different pseudoconcave utility functions. The independent parameters of the experimental study are as follows: CC (Options on Cut constraints), SPS (Solution space Partitioning Strategies), IC (options on setting up Initial convex Cones), SA (Search Algorithms) and STS (Search Termination Strategies). The levels of these parameters are: CC (Yes-No: whether to use cuts or not), SPS (N, SPS1, SPS2: No partitioning, Partitioning by methods SPS1 or SPS2), IC (IC1, IC2: setting up convex cones using methods IC1 or IC2), SA (SA1, SA2, SA3: Search Algorithms SA1, SA2 or SA3, where SA1 and SA2 are discussed in Ramesh *et al.* [1986] and SA3 is a hybrid of the two strategies), STS (STS1, STS2, STS3: STS1 and STS2 are heuristics and STS3 is a rigorous procedure). A completely randomized block design consists of 180 independent cells. Therefore, even if a sample of five problems per cell is used, it will require solving 900 problems to complete the design. Furthermore, if the design is repeated for different utility functions, then the number of problems to be solved becomes extremely large. Therefore, a sequential design scheme to evaluate the main effects of the parameters under specific conditions is used. This scheme has been derived from our initial experience from the preliminary study of the solution framework. Accordingly, the following experimental designs have been constructed.

In *Design 1*, the following parameters are fixed as follows: CC: Yes, SPS: SPS1, IC: IC1, and STS: STS1. In this design, the parameter SA is varied. The computational results of Design 1 showed that search algorithm SA3 performs better than the other two algorithms in terms of both the number of questions and solution time than the other algorithms. Therefore, using SA3 as the search algorithm, and the same levels of CC, SPS and IC, the search termination strategies are evaluated in *Design 2.* The heuristic STS1 obtained the optimal solution in all the problems solved in Design 2. Therefore, using STS1 and the same levels of CC, IC and SA, the set partitioning strategies are evaluated in *Design 3.* The strategy SPS1 resulted in fewer number of questions in this design. Therefore, using SPS1 and the same levels of IC, SA and STS, the effects of enforcing the cut constraints on the objectives are evaluated in *Design 4.* Enforcing the cuts resulted in significantly reducing the number of questions. Therefore, enforcing the cut constraints with the same levels of SPS, SA and STS, the effects of the options on setting up initial convex cones are investigated in *Design 5.* The method IC1 resulted in fewer questions than the method IC2. Therefore, using IC1 and the same levels of CC, SPS, SA and STS, a completely randomized block design study of the solution framework for different problem sizes and utility functions is carried out in *Design 6.* The problem size parameter in Design 6 consists of three levels

Table II: Performance Statistics: Design 2 †

Problem size	# Questions for (BCIP)			Sol. Time for (BCIP)		
	STS1	STS2	STS3	STS1	STS2	STS3
(20 × 40)	16.74	17.54	24.91	7.64	11.21	24.54
(30 × 70)	17.85	20.45	27.22	8.90	12.72	24.45

† Each cell is the average of twenty problems.

Table III: Performance Statistics: Design 3 †

Problem size	# Questions for (BCIP)			Sol. Time for (BCIP)		
	SPS1	SPS2	NONE	SPS1	SPS2	NONE
(20 × 40)	16.74	17.22	27.41	7.64	8.32	15.25
(30 × 70)	17.85	17.44	31.16	8.90	9.54	22.67

† Each cell is the average of twenty problems.

in number of variables (20, 50, 100) and three levels in number of constraints (10, 30, 50). Four utility functions have been tested with each problem size. These are: quadratic with weights (0.3, 0.7), quadratic with (0.5, 0.5), fourth degree polynomial with weights (0.3, 0.7) and fourth degree polynomial with weights (0.5, 0.5). We summarize these results in the following discussion.

Design 1

The performance characteristics of the three search algorithms for two representative problem sizes are provided in Table I. Algorithm SA3 requires the smallest number of questions. Further, algorithm SA1 requires significantly more number of questions than SA2. In algorithm SA1, branching from a noninteger is solution is continued until an integer solution is obtained before performing the Zionts and Wallenius procedure, while this procedure is performed at every candidate problem in SA2. These results show that algorithm SA1 could entail considerable branching. Algorithm SA3 which is a hybrid of the other two strategies and algorithm SA2 show significantly better performance than algorithm SA1. The ratio of the number of questions required to solve problem (BCIP) to that of problem (BCLP) is significant for all the three algorithms. Similar behavior can be observed with solution times too. Thus, problem (BCIP) is more difficult to solve than problem (BCLP) by an order of magnitude. In general, the percentage of questions saved using the two fathoming tests is significant, and resulted in a savings between 15

Design 2

Table II summarizes the results from this study. The average number of questions on pairwise comparisons is the smallest when the heuristic STS1 is used. The average number of tradeoff questions is less with STS2 than with the rigorous procedure STS3. This behavior is expected. Further, the two heuristics obtained the optimal integer solutions in all the problems tested. This shows that the heuristics, especially STS1, are fairly robust and efficient in solving problems of the class tested.

Design 3

Table III summarizes the results from this study. Partitioning the solution space according to either SPS1 or SPS2 results in fewer questions than when no partitioning is used. The preliminary computational results with SA1 showed even more substantial improvement in the number of questions when the solution space is partitioned. This is because, SA1 generates many dominated solutions in the branching process, and their consideration is restricted by partitioning the solution space. There is no significant difference between the partitioning strategies in terms of solution time. However, the number of candidate problems increases when the partitioning strategies are used. This is because the integer programming problem is solved separately in each partition.

Design 4

Enforcing the cut constraints resulted in saving 20% to 27% of the number of questions required to solve problem (BCIP) in the class of problems solved. These statistics show that enforcing the cut constraints results in fewer questions. However, there is no significant change in the solution time. Although enforcing the cuts and partitioning the solution space lead to problems of degeneracy in the solution tableau (see Ramesh *et al.* [1990]), they are preferable since the total number of questions can be significantly reduced.

Design 5

This study did not yield any appreciable difference in the number of questions required with methods IC1 and IC2. However, Method IC2 requires a few more questions than method IC1 in setting up the initial convex cones. On the average, this minimal difference did not make a significant difference between the two methods in terms of the questions required in solving problem (BCIP). Further, there does not appear to be any appreciable difference in the solution times between the two methods as well.

Design 6

We summarize some of the key results from this study in the following discussion. The number of questions increases with the number of variables for the four utility functions tested. This trend is observed with all the search termination strategies. Since the number of efficient solutions increases with the number of variables, additional exploration of the decision-maker's preference structure is necessary. In the case of both the utility functions with weights (0.3, 0.7), the rate of increase decreases with the number of variables. However, in the case of the utility functions with weights (0.5, 0.5), such a consistent behavior is not observed. This suggests possible effects of interaction between the number of variables and number of constraints in this case. However, the number of questions does not show an increasing trend with the number of constraints with all the utility functions tested. The effects of an increase in the number of constraints is twofold. It could restrict the set of feasible solutions to the problem, while also increasing the set of efficient solutions. Accordingly, the effects of the number of constraints on the number of questions have been fluctuating, and could depend to a large extent on the geometry of the polyhedral sets. The weights ((.3, .7) on the objectives yield elliptical utility contours in the quadratic case. Similarly, the weights (0.5, 0.5) yield circular contours of the quadratic utility function. The computational results show that the curvatures of the utility contours may have a significant effect on the number of questions, with the more elliptic contours requiring less number of questions. Compared to the behavior of the number of questions, the solutions times behave more consistently. The solution time increases with both the number of variables and the number of constraints. Furthermore, unlike in the case of number of questions, the rate of increase in the solution time increases with the number of variables. Among the four types of utility functions considered, there does not appear to be any significant difference in terms of solution time.

4 Conclusion

In this research, we have developed an efficient interactive solution framework for bicriteria integer programming. This framework has been developed for the large class of decision makers with underlying pseudoconcave nondecreasing utility functions. This framework consists of six algorithmic components. The theory underlying these components has been presented in our earlier articles, and we present results from a detailed empirical investigation of the performance of these components in this article. The framework has been extensively tested for various options on the procedural elements of the solution approach, and we conclude that using the set of options {CC: Y, SPS: SPS1, IC: IC1, SA: SA3, STS: STS1} for solving problems of the class tested will be most appropriate from this study. This framework can also be extended to solve multicriteria integer programming problems in general, and we are currently working in this area. Some of the important applications of bicriteria integer programming arise in quality control, university examination scheduling, portfolio selection, warehouse location, strategic planning and policy analysis, and many more. We are also currently investigating an important application in the area of systems design using the proposed solution approach.

References

[1] Benson, H. P., "Vector Maximization with Two Objective Functions," *Journal of Optimization Theory and Applications*, Vol. 28, 1979.

[2] Chalmet, L. G., Lemonidis, L. and Elzinga, D. J., "An Algorithm for the Bicriterion Integer Programming Problems," *European Journal of Operations Research*, Vol. 25, 1986.

[3] Evans, G. W., "An Overview of Techniques for Solving Multiobjective Mathematical Programs," *Management Science*, Vol. 30, No. 11, 1984.

[4] Gabbani, D. and Magazine, M. J., "An Interactive Heuristic Approach for Multiobjective Integer Programming Problems," *Journal of the Operational Research Society*, Vol. 37, 1986.

[5] Geoffrion, A. M., "Solving Bicriterion Mathematical Programs," *Operations Research*, Vol. 15, 1967.

[6] Marcotte, O. and Soland, R. M., "An Interactive Branch-and-Bound Algorithm for Multiple Criteria Optimization," *Management Science*, Vol. 32, No. 1, 1986.

[7] Pasternak, H. and Passy, U., "Bicriterion Mathematical Programs with Boolean Variables," in *Multiple Criteria Decision Making*, J. L. Cochrane and M. Zeleny (Eds.), University of South Carolina Press, Columbia, SC.,1982.

[8] Ramesh, R., Karwan, M. H. and Zionts, S., "An Interactive Method for Bicriteria Integer Programming," *IEEE Transactions on Systems, Man and Cybernetics*, Vol. 20, No. 3, 1990.

[9] Ramesh, R., Karwan, M. H. and Zionts, S., "Preference Structure Representation Using Convex Cones in Multicriteria Integer Programming," *Management Science*, Vol. 35, No. 9, 1989.

[10] Ramesh, R., Karwan, M. H. and Zionts, S., "A Class of Practical Interactive Branch and Bound Algorithms for Multicriteria Integer Programming," *European Journal of Operations Research*, Vol. 26, No.1, 1986.

[11] Zionts, S. and Wallenius, J., "An Interactive Multiple Objective Linear Programming Method for a Class of Underlying Nonlinear Utility Functions," *Management Science*, Vol. 29, No. 5, 1983.

INTERACTIVE APPROACHES FOR SOLVING SOME DECISION MAKING PROBLEMS IN THE CZECHOSLOVAK POWER INDUSTRY

Dr. Martin Černý and
Dr.Dagmar Glückaufová
Institute of Economics
Czechoslovak Academy of Sciences
Politickych veznu 7
111 73 Prague 1
Czechoslovakia

Abstract: In the paper some interactive approaches to solving multicriteria decision making problems are summarized together with their applications to selected problems of the Czechoslovak power industry. The methodological tools applied are the sensitivity analysis of parameters in the Electra I method based on fuzzy preference relations and the interactive STEM vector optimization method where the possibility of allowing fuzzy answers by the decision maker is also considered.

Keywords: Multicriteria decision making, interaction, fuzzy preference relation, MADM methods, Electra I, vector optimization, STEM method, fuzzy optimization.

1. INTRODUCTION

Many models of decision making situations depart from the assumption that the decision-maker's preferences and requirements can be specified at least in principle in advance. In practical situations, however, such specification often appears to be very intricate and sometimes even an impossible task. Such problems cannot be solved separately without communication with a decision-maker. Instead, exchange of information between a decision-maker and an analyst is needed.

Such a conversation between the analyst and decision-maker may concern the formulation of a decision model (e.g., the choice od criteria, generation of a set of alternatives, changes in constraints, etc.). In such a case we speak about interactions in a broader sense. Alternatively, interactions may form a necessary component of the process of solving an already formulated model. Such interaction is called an interaction in a narrower sense (e.g., the determination of threshold values, gradual establishement of a decision-maker's preference structure etc.) - see, e.g., Bouska, Cerny, Glückaufová (1984). In both cases such a conversation requires repeated computations to be performed; using the results obtained the decision-maker modifies the input data for the next computation.

In the modelling of decision-making situations, we are faced in all the stages of the process with incomplete information which is available. Various causes may be responsible for this lack if information. On the one hand, there are situations when we do not know which of the anticipated events will occur in the future and what effect will they exert on the result of the decision-making process. In this case, we speak of decision making under risk and uncertainty.

On the other hand, there are situations when we are unable to strictly define the circumstances anticipated, when we have only hazy, vague ideas of these circumstances. Here, we speak of decision making under indeterminancy.

The basic mathematical tool for decision-making under risk and uncertainty is the statistical theory of decision making based on the conncept of probability. The corresponding mathematical apparatus has already been well developed and elaborated. The modelling of specific aspects of indeterminancy presents, from the mathematical point of view, a far more difficult problem, mainly because of the fact that it is necessary to represent, by means of exact instruments, vaguely defined notions, both with respect to decision-making alternatives and with respect to the decision maker's objectives and their interrelations. To handle those problems the fuzzy sets approach las proved to be extremely useful.

In this paper we shall deal with a few decision making problems under indeterminancy, with the help of the above mentioned apparatus. All those problems came from power generating industry. The problems solved were of two different types:

- the problems with decision alternatives explicitly given,
- the problems where the alternatives were generated during the process of solving the problem.

The methods developped for solving the problems of the first type are commonly known as the multiaspect decision making (MADM) methods. The problems of the second type are solved with the help of multicriteria programming or vector optimization methods.

2. APPLICATION OF MADM METHODS

2. 1. Outline of the applied methodology

A significant common feature of the MADM methods is the presence of subjective factors represented by parameters the values of which are given by the decision-maker. Such parameters are in the first place the weights expressing a relative importance of criteria which form the essential input in most methods. Apart from that many MADM methods require the specification of the so called threshold values which are not always easy to specify owing to their difficult interpretation.

In this context it is obvious why sensitivity analysis plays an important role in solving practical problems. In other words we are interested in the problem of under which changes of the values of exogeneous parameters the particular evaluation remains unchanged.

There exist a wide variety of the MADM methods, among which the Electra class of methods are of great importance mainly because of their relative transparency and applicability the sinplest of those methods is Electra I which is particularly suitable to deal with in vague decision making situations for two reasons.

First, it provides, instead of ordering the whole set of alternatives, only a subset of nondominated alternatives which is appropriate to get from indeterminated and vague data.

Second in the case of such simple method sensitivity analysis can be easily performed and the conditions under which the set of nondominated alternatives can be studied.

Therefore we have limited ourselves to the analysis of stability of the Electra I method.

To present some ideas about the approaches used it is necessary to give at least a short summary of the Electra I method as well as of the principles of sensitivity analysis applied in this case.

2.1.1. The Electra I Method

Let $X = \{x_1, \dots, x_n\}$ denote a finite set of alternatives. The alternatives are evaluated in terms of m quantitative criteria. The evaluation can be summarized by a matrix $B = (b_{ij})$, $i = 1, \dots, n$, $j = 1, \dots, m$, where b_{ij} is the value of the jth criterion for the ith alternative. Further let $p_1, \dots, p_m$ be weights expressing a relative importance of the criteria. The data matrix B will be normalized as follows:

$$u_{ij} = \frac{b_{ij}}{\sum_{k=1}^{n} b^2_{kj}} \qquad (i=1,2,\dots,n;\ j=1,2,\dots,m).$$

The outranking relation R can be represented by an incidence matrix $R = (r_{ij})$. If an alternative x_i has to be preferred to x_j according to the outranking relation, the following two requirements should be fulfilled:

- x_i has to be better than x_j subject to a sufficient number of sufficiently important criteria,

- x_i must not be much worse than x_j in the remaining criteria.

To meet these requirements two fuzzy relations on the set X are constructed. The first of them is called concordance relation and is represented by a membership matrix $C = (c_{ij})$, where

$$c_{ij} = \sum_{k \in C_{ij}} p_k,$$

$$C_{ij} = \{k / u_{ik} \geq u_{jk}\}.$$

The second of them, the discordance relation has a membership matrix $D = (d_{ij})$ where

$$d_{ij} = \frac{\max\limits_{k \notin C_{ij}} (u_{jk} - u_{ij}) p_k}{\max |u_{jk} - u_{ij}| p_k}$$

The decision maker is asked to provide two parameters, a concordance threshold c and a discordance threshold d. Both values are chosen from the <0,1> interval.

The incidence matrix $R = (r_{ij})$ of a crisp outranking relation is then defined as follows

$$r_{ij} = \begin{cases} 1 \text{ if } c_{ij} \geq c \text{ and } d_{ij} \leq d, \\ 0 \text{ otherwise} \end{cases}$$

On the basis of R we define a set of nondominated alternatives E(R):

$$E(R) = \{x_j \mid r_{ji} = 0, \text{ for all } j\}.$$

2.1.2. Sensitivity analysis for Electra I

The sensitivity analysis can be carried out for the values of thresholds c and d as well as for the values of weights p_j, $j=1,\dots,m$.

The essence of sensitivity analysis (see Vetschera, 1986) lies in the construction of intervals of stability, i.e. intervals for the values of parameters mentioned above for which the set of nondominanted alternatives E(R) does not change. The interval for each particular parameter is derived under the assumption that the values of all other parameters remain unchanged. For example, a stability interval for the weight of one attribute is found, provided that all other weights (as well as the threshold values) do not change. Apart from the numerical values for the limits of a stability interval, stability analysis can also find out what happens when those limits are reached, i.e. which alternative enters or leaves the set E(R).

Sensitivity analysis can be utilized in two different ways. On the one hand, it is possible to construct stability intervals for weights and to answer the user's questions concerning the influence of the weights on the solution obtained. For example, while solving practical problems we were often asked by the users to what extent the criteria, or more precisely their weights, influence the final ordering of alternatives.

On the other hand sensitivity analysis, especially stability intervals for the threshold parameters can be used in an interactive way for the reduction of the set of nondominated alternatives.

2. 2. Practical application

Both ways of using stability analysis were applied to some practical problems in the field of the production and use of energy. The problem we shall present here concerned the evaluation of alternative heat sources from various economical as well as ecological aspects (see Glückaufova at al, 1988). The following 8 heat sources were considered:

1. Centralized production from lignite coal (BCC)
2. Decentralized production from natural gas (NGD)
3. Centralized production from natural gas (NGC)
4. Decentralized production from liquid fuels (LFD)
5. Centralized production from liquid fuels (LFC)
6. Centralized production in nuclear power plants (NPC)
7. Decentralized production from electrical energy (EED)
8. Combined alternative 6 and 7 (NPC + EED)

Those alternatives were evaluated according to 9 criteria representing different relevant aspects. In the first step we used the Electra I method with equal weights. The set of nondominated alternatives consisted of there alternatives:
E(R) = {NGD, NGC, NPC}.

In sensitivity analysis concerning the weights we have concentrated on the criteria according to which the nondominated alternatives were evaluated as worse than the others. From the criteria with such a property the most outstanding was the criterion of heat transfer costs. For the special case of this criterion we have obtained following results:

- the alternative BCC can never enter the set E(R),
- if the weight increases from the initial value 0.11 to 0.59, alternative NGD leaves the set E(R),
- by the change of the weight it is possible to exclude alternative NGC from the set E(R),
- by increasing the weight the first alternative to enter the set E(R) is alternative NPC + EED.

The second methodological approach mentioned in the previous subsection, the successive reduction of the set E(R), was applied to the same practical problem. The procedure starts with the most demanding values of thresholds c and d. In our case we started with c = 0.957 and d = 0.118. For those values the outranking relation R is empty so that all alternatives form the set E(R). The successive elimination of alternatives through the weakening of thresholds proceeded in the steps summarized in the Table 1.

Table 1

Step No	Value of c	Value of d	Alternative leaving E(R)
1	0,957	0,300	NGC
2	0,746	0,300	NPC
3	0,682	0,300	NPC + EED
4	0,561	0,300	NGD
5	0,561	0,476	LFD
6	0,529	0,476	LFC
7	0,529	0,668	EED

As a result the set of nondominated alternatives was reduced to a single element BCC.

3. APPLICATION OF INTERACTIVE VECTOR OPTIMIZATION

To solve problems when the alternatives are generated during the process of solving the problem we have used interactive vector optimization approach as we needed to maintain a contact with the decision maker during the whole process of finding a solution.

From a wide variety od interactive methods we have chosen the STEM method because this method is both quite transparent to the decision maker as well as applicable to a wide range of vector optimization problems.

3. 1. Short description of modified STEM method

We shall now describe a modification of the STEM method which we have developped and practically implemented. Generally, the method can be used to solve any problem of the following form:

$(f_1(x),\dots,f_m(x))$ "max"

s.t. $x \in X$.

Here $f_k(x)$, $k=1,\dots,m$, are the objective functions and is X a feasible set (decision set) of the problem.

At present our algorithmic and programming support allows us to solve the problems of two types:

a) linear vector optimization problems,

b) problems with a finite set of alternatives, i.e. problems where X is a finite set (which may be of quite a large size) and the values of the objective functions are given in the matrix form.

The algorithm starts by solving m partial optimization problems with respect to individual objective functions.

Then we form a partial optima matrix (pay-off matrix) (z_{ik}), $i,k = 1,2,\dots,m$, $z_{ik} = f_k(x^{*(i)})$ is the value of ith objective attained at the k-th partial optimum.

The values on the principal diagonal of this matrix represent the maximal attainable values (ideal values) of the objective functions on the feasible set X. Let us write

$z^*_k = z_{kk} = f_k(x^{*(k)})$ $(k = 1,2,\dots,m)$

and let

$z^* = (z^*_1,z^*_2,\dots,z^*_m) \in R_m$

be the so called ideal vector (or ideal point).

In the q-th step of the algorithm $(q = 1,2,\dots)$ the following problem is solved:

$$\max_{k \in K^{(q)}_2} \frac{1}{u_k}[z^*_k - f_k(x)] \quad \min$$

subject to $x \in X$, $f_k(x) \geq h^{(q)}_k$, $k \in K^{(q)}_1$,

where:

u_k - the unit of measuring the deviation of the k-th objective function from its ideal value (the u_k's are given by the decision maker),

$K^{(q)}_1$ - the set of indices of those objective functions the values of which (as they were reached in the previous step) are judged by the decision-maker as satisfactory (for q = 1 we set $K^{(1)}_1 = \emptyset$),

$K^{(q)}_2 = \{1,2,\dots,m\}$ $K^{(q)}_1$ - the set of indices of those objective functions the values of which are not judged satisfactory by the decision-maker,

$h^{(q)}_k$ - the satisfaction level of the k-th objective function given by the decision-maker usually in the form

$h^{(q)}_k = f_k(x^{(q-1)}) - \Delta_k$,

where $f_k(x^{(q-1)}$ is the value of the k-th objective function reached in the previous (q-1)-th step and Δ_k is the maximum relaxation which the decision-maker can tolerate.

As we have seen the information required from the decision-maker by the modified

STEM algorithm at each iteration step consists of

- selecting objective functions whose levels are already satisfactory (i.e. the set $K^{(q)}{}_1$),
- determining satisfaction limits $h^{(q)}{}_k$ or the tolerances ${}_k$ for those objective functions ($k \in K^{(q)}{}_1$).

In some situations it may be difficult for the decision-maker to give exact answers to the questions put by the algorithm. This is true particularly in the case when there is a collective body (committee) of decision-makers whose preferences may differ. It may therefore be useful to allow for fuzzy answers by the decision-maker. At present we have extended the modified STEM algorithm described above so that it can handle fuzzy satisfaction limits instead of exact ones. The problem then becomes a fuzzy optimization one which is formulated according to the Bellman-Zadeh approach. See Cerny, Gluckaufova (1987).

3. 2. Application to a practical problem

The modified STEM method described above in its nonfuzzy version was used to solve a problem concerning the choice of the mode of utilization of existing coal power plants in Northern Bohemia. Originally the problem was formulated as a bivalent one; as we are at present not able to handle such multicriterion problems directly, we have solved the problem in two phases:

In the first phase we have generated a sufficient number of feasible solutions of the problem using a specially developped enumeration algorithm.

In the second phase we have found compromise solution using the STEM method with a finite set of alternatives, where the feasible set of alternatives was generated in the previous phase.

Later on, on the basis of the experience gained by solving the problem in its bivalent form, the problem was reformulated as a mixed-integer multiobjective linear programming one. The decision variables were the numbers of power-generating units subject to alternative ways of treatment, namely:

- leave the unit as it is (without reconstructing),
- install a smoke-filtering equipment (two alternative types were considered),
- stop the work of the unit and replace its capacity by a nuclear power plant.

The levels of production of electricity by the units (with or without reconstruction) were to be determined as well.

The constraints of the problem represent the required final eletricity output (which can be lowered, at some cost and to certain extent, by consumption-reducing measures adopted by the consumers) and also the limits on the capacity of nuclear power plants replacing the scrapped units.

The objective functions of the problem were as follows:

- investment costs (10^9 Kcs) - C,
- emissions of sulphurous oxide (10^3 t per year) - E,
- running costs (10^6 Kcs per year) - V,
- fuel consumption (10^3 t per year) - P.

All objective functions are of course to be minimized.

The mixed-integer linear programming problem obtained as a result was solved for the practical case of 16 power generating units of 100/110 MW capacity and of 16 units of 200/210 MW capacity. The required electricity production was considered to be 22000 GWh per year (or by 500 GWh less when power-saving measures are adopted by the customers); maximum capacity of replacing nuclear power plants was set to 1000 MW.

The problem was solved by the modified STEM method adapted for solving mixed-integer problems. As the first step, the payoff matrix was determined in Table 2:

Table 2

Minimized objective	Value of C	E	V	P
C	0	735,3	4792	236,0
E	33,57	64,2	7175	204,7
V	0,75	722,1	4782	231,6
P	13,65	495,9	4789	158,4

After an analysis of the matrix, the decision-maker has formulated the following two particular problems to solve:

a) What are the minimum investment cost with which the emissions of sulphurous oxide do not exceed 500.10^3 t per year?

b) For several different levels of investment cost, what are the minimum running costs, provided that the emissions are sufficiently low?

To find an answer to question a), we first set the measuring units u_i (which were originally all equal to 1) in such a way, so as to compensate for the different ranges of individual objectives values; i.e.: $u_1 = 0{,}1$; $u_2 = u_4 = 1$; $u_3 = 10$. Then we applied the STEM method and in three steps we got the solution with the objective values:

$C = 0{,}97$; $E = 500$; $V = 5515$; $P = 246{,}8$

with the given limits

$k_E = 500$; $k_V = 7218$; $k_P = 413{,}8$.

As the limits for objectives V and P are not reached it is clear that the value $C = 0{,}97$ represents the minimum investment level.

Question b) was dealt with for the following specified levels of investment C:

C = 5; 10; 13,5; 20; 25.

For each of those levels a solution representing some compromise between other three objectives was found; then the decision-maker determined the acceptable level of emissions E and under those conditions the relation between the two remaining objectives was analyzed. For example, when C = 13,5, first compromise solution was found giving the values

C = 13,5; E = 237,9; V = 6706; P = 258,5.

The decision-maker then set the limit of E = 300, which led to the values

C = 13,5; E = 300; V = 6130; P = 232,7.

As the value of V was still rather large, the decision-maker raised the acceptable level of E to 350 and got the values

C = 13,5; E = 350; V = 5642; P = 198,7,

which was considered satisfactory. The corresponding compromise solution was therefore adopted as "best" for the investment level C = 13,5. Similar analysis was carried out for all other levels of investment cost.

REFERENCES

Bouška J., Černý M., Glückaufová D. (1984) Interaktivní postupy rozhodování (Interactive Decision Making Procedures - in Czech). Academia, Prague.

Černý M., Glückaufová D. (1987) Fuzzy Concepts in Multiaspect Decision Making. Ekonomicko-matematický obzor 23, 55-65.

Glückaufová D., Fapšo O., Spitz J. (1988) Sensitivity Analysis of Solutions of a MCDM Problem - Case of Energetics (in Czech). Ekonomicko-matematicky obzor 24, 405-416.

Vetschera R. (1986) Sensitivity Analysis for the Electra Multicriteria Method. Opns. Res. 30, 99-117.

Vetschera R. (1988) An Interactive Outranking System for Multiattribute Decision Making. Comput. Opns. Res. 15, 311-322.

3. ISSUES RELATED TO INTERACTIVE DECISION MAKING

ELICITATION OF OPINIONS BY MEANS OF POSSIBILISTIC SEQUENCES OF QUESTIONS

Bernadette Bouchon-Meunier
CNRS
LAFORIA, Université Paris VI, Tour 46
4 place Jussieu, 75252 Paris Cédex 05, France

Abstract : Sequences of questions are used to elicit preferences of a given population in regards to a particular decision to make. The answers obtained from the inquired persons are allowed to be slightly different from the expected opinions, expressed by linguistic labels or weighted with coefficients indicating an uncertainty. We show how to interpret these answers with regard to the decision, and we also consider the case where the expected opinions are indicatory of the decision rather than absolutely conclusive. In this case, we propose a method to determine the sequence of questions providing the most efficient information with regard to the decision to make.

Keywords : Preference elicitation, questions, possibility theory, uncertainty.

1. INTRODUCTION

Asking questions is a classical method to elicit the preferences of a given population with regard to a particular decision to make. The answers seem sometimes difficult to express and the inquired persons may hesitate when choosing a particular state **s** of a criterion or a decision. In this case, it can be useful to ask them to weight their answers with coefficients or to use linguistic labels indicating to what extent **s** is the best answer from their point of view.

Questionnaires can be studied in a framework of probabilistic uncertainty (Bouchon, 1988) and they can be defined in a fuzzy setting, the answers to the questions being defined as fuzzy subsets of a universe of description (Bouchon, 1985). Possibility and necessity measures can also be used to represent the doubts of the persons (Bouchon, 1990). The elicitation may be easier when more flexibility is allowed in the answers, or when the linguistic expression of the responses can be freely chosen. Coefficients associated with the elements of the universe in question either quantify the degree to which a chosen issue of a query belongs to the effective opinion expressed by the inquired person, or represent the credibility the inquirer assigns to the person, or define the preference scale characterizing a given criteria (Bouchon, 1987).

Instead of being a handicap for the decision-making, the uncertainties of the population are taken into account as an important element of the answers for the search after a decision satisfying the preferences of the population to the best.

The construction of sequences of questions yielding the elicitation of opinions leading to a final decision is then associated with the management of the uncertainty expressed by the population. In this paper, questionnaires are manipulated in a possibilistic framework and an interface between symbolic and numerical answers is possible.

2. PREFERENCE ELICITATION BY MEANS OF SEQUENCES OF QUESTIONS

Let us consider a given set $D = \{d_1, ..., d_n\}$ of decisions to identify, which are supposed to be defined without any ambiguity.

We also consider a set $Q = \{q_1, ..., q_m\}$ of questions, where q_i is a link between a linguistic variable X_i defined on a universe U_i and a family of a(i) labels, denoted by q_i^1, ..., $q_i^{a(i)}$, and associated with possibility distributions π_i^1, ..., $\pi_i^{a(i)}$, defined on U_i and lying in [0, 1].

The inquired persons are allowed to provide answers in a way somewhat different from the expression we expect, and we denote by q'_i the label obtained as an answer to question q_i, $1 \leq i \leq m$, more or less different from all the q_i^k, $1 \leq k \leq a(i)$, and by π'_i the possibility distribution describing q'_i, defined on U_i and lying in [0, 1] (see Figure 1).

We can further suppose that the relationship between the obtained answers to questions are not sufficient to strictly determine a unique decision to make and that they are only indicatory of the suitability of the possible decisions for the expressed opinions.

The sequence of questions to be put to the population can be either constructed by means of a possibilistic method taking into account the knowledge provided by every question about the possible decisions (case 1), or determined in a prior way by the inquirers, according to the meaning of the answers with respect to the decision to make (case 2).

In case 1, we successively choose a question in Q to improve the certainty we have about the suitability of a decision of D. We describe in section 4 a method to construct such a questionnaire.

Let us suppose given the possibility $\pi(d_j / q_i^k)$ that the decision d_j of D is satisfying, for $1 \leq j \leq n$, when we obtain the answer q_i^k to question q_i, for $1 \leq i \leq m$, $1 \leq k \leq a(i)$. We also suppose given the necessity $N(d_j / q_i^k)$ quantifying the certainty that this answer implies that d_j must be identified. Both possibility and necessity coefficients belong to [0, 1] and they are such that $N(d_j / q_i^k) \leq \pi(d_j / q_i^k)$.

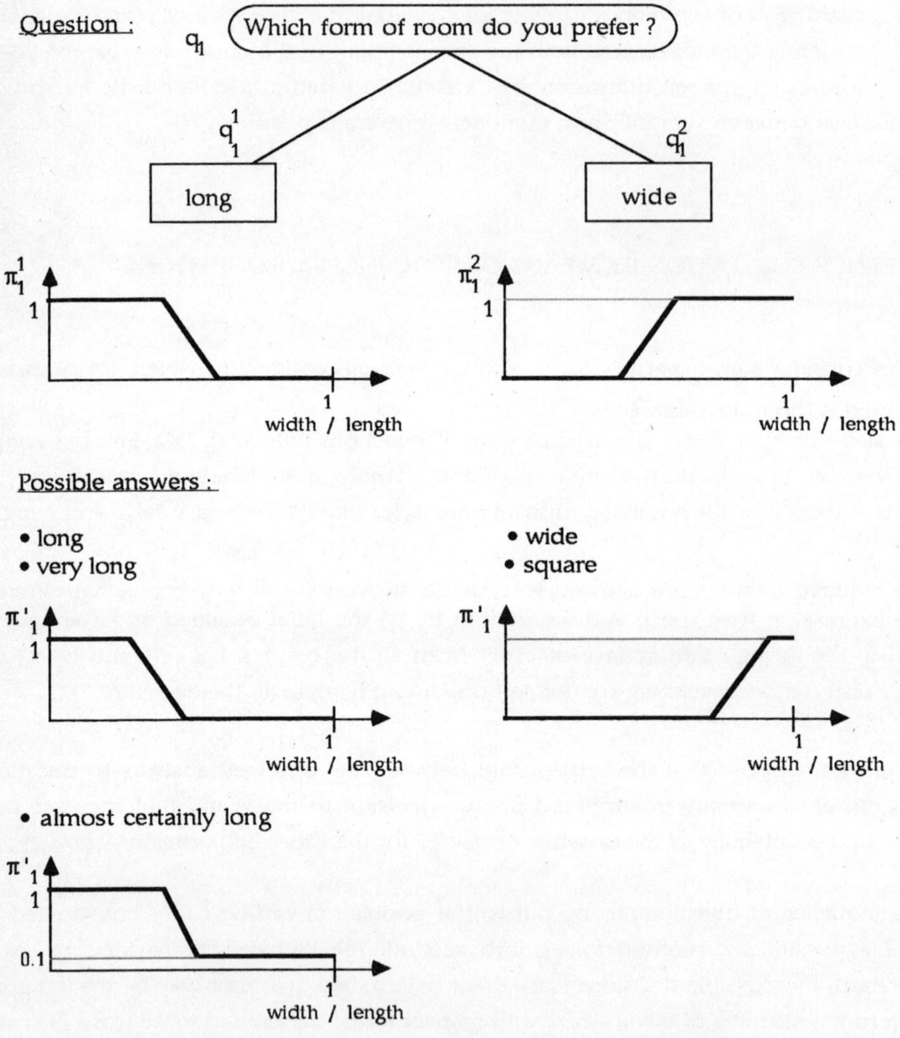

Figure 1 : example of question and possible answers

We can also suppose given some knowledge about the fact that decision d_j can be chosen, when an answer different from q_i^k is obtained to question q_i : let $\pi(d_j / \neg q_i^k)$ and $N(d_j / \neg q_i^k)$ denote the possibility and the necessity that d_j is acceptable when q_i^k is not obtained. If the values of π and N are not precisely known, the interval to which they belong, generally [0, 1] itself, will be used (Farreny et al., 1986).

If q_i^k is determinative for d_j, then $N(d_j / q_i^k) = \pi(d_j / q_i^k) = 1$. Otherwise, a sequence $x_1, ..., x_r$ of questions is necessary to determine an element d_{jo} of D such that its possibility to be satisfying, given the answers provided to questions $x_1, ..., x_r$, is sufficiently high and the certainty available on its identification is acceptable. If $x_1^{k(1)}, ..., x_r^{k(r)}$ are answers to questions $x_1, ..., x_r$, a possibility coefficient and a necessity coefficient weight the identification of every decision d_j and they are evaluated as :

$$Pos\ (d_j / x_1^{k(1)}, ..., x_r^{k(r)}) = \min_{1 \leq i \leq r} \pi(d_j / x_i^{k(i)}),$$

$$Nec\ (d_j / x_1^{k(1)}, ..., x_r^{k(r)}) = \max_{1 \leq i \leq r} N(d_j / x_i^{k(i)}).$$

In case 2, the values of $Pos\ (d_j / x_1^{k(1)}, ..., x_r^{k(r)})$ and $Nec\ (d_j / x_1^{k(1)}, ..., x_r^{k(r)})$, corresponding to the admissibility of a decision for a given sequence of answers $x_1^{k(1)}, ..., x_r^{k(r)}$ to questions $x_1, ..., x_r$ chosen in Q, is given by the inquirers without considering the elementary values $\pi\ (d_j / q_i^k)$ associated with every answer to a question.

In both cases, the decision d_{jo} will be preferred if $Pos\ (d_{jo} / x_1^{k(1)}, ..., x_r^{k(r)})$ is the highest value of $Pos\ (d_j / x_1^{k(1)}, ..., x_r^{k(r)})$, for all the d_j in D. In case of ambiguity, the highest value of $Nec\ (d_j / x_1^{k(1)}, ..., x_r^{k(r)})$ will be chosen.

We further suppose that we look for a decision satisfying a given population of p persons and we ask them questions of Q in order to identify their opinion.

3. ADEQUACY OF AN ANSWER TO THE EXPECTED SCHEME

Let us consider an inquired person IP answering a sequence of questions. An element of D can be identified from the answers to the various questions. Let $x_1, ..., x_r$ denote a sequence of questions of Q in the questionnaire, such that IP provides answers $x_1^{k(1)}, ..., x_r^{k(r)}$ to these questions identical with the expected labels. Then, the most suitable decision to IP is d_{jo} such that $Pos\ (d_{jo} / x_1^{k(1)}, ..., x_r^{k(r)})$ is the highest value, as well as $Nec\ (d_{jo} / x_1^{k(1)}, ..., x_r^{k(r)})$ in case of several solutions.

If IP provides answers different from the expected labels, the resemblance of the answer q'_i given by IP with one of the labels proposed by q_i is measured by the following adequation coefficients (Dubois and Prade, 1987):

- a possibility measure of the adequation of the answer with a given label :

$$\pi(q_i^k; q'_i) = \sup_{\{u \text{ in } U_i\}} \min\ (\pi_i^k(u), \pi'_i(u)),\ 1 \leq k \leq a(i),$$

- a necessity measure of this adequation :

$$N(q_i^k; q'_i) = \inf_{\{u \text{ in } U_i\}} \max\ (1 - \pi_i^k(u), \pi'_i(u)),\ 1 \leq k \leq a(i).$$

We shall consider the closest label q_i^k to q'_i, corresponding to the greatest value of $\pi(q_i^k; q'_i)$ and in case of ambiguity, to the greatest value of $N(q_i^k; q'_i)$.

The possibility that any d_j is convenient to IP, and the certainty of this assertion, will be evaluated by the following calculus (Wyss, 1988) :

$$\pi^k(d_j)=\max[\min\{\pi(d_j / q_i^k),\pi(q_i^k;q'_i)\},\min\{\pi(d_j / \neg q_i^k),1-N(q_i^k;q'_i)\}],$$

$$N^k(d_j)=\min[\max\{N(d_j / q_i^k),1-\pi(q_i^k;q'_i)\},\max\{N(d_j / \neg q_i^k),\ N(q_i^k;q'_i)\}].$$

As indicated in Farreny et al. (1986), the values can be replaced by the interval to which they belong in the case where they are not precisely known.

Now, let us consider again a sequence of questions $x_1, \ldots, x_r$. Because of the differences which may exist between the expected answers to these questions, and the labels obtained from IP, the possibility and certainty of any decision d_j of D will be the following :

$$\mathrm{Pos}\ (d_j\ /\ x_1^{k(1)}, \ldots, x_r^{k(r)}) = \min_{1\leq i\leq r} \pi^{k(i)}(d_j),$$

$$\mathrm{Nec}\ (d_j\ /\ x_1^{k(1)}, \ldots, x_r^{k(r)}) = \max_{1\leq i\leq r} N^{k(i)}(d_j).$$

The element d_{jo} will be definitely identified by means of the sequence of questions x_1, ..., x_r if there exists labels $x_1^{k(1)}, \ldots, x_r^{k(r)}$ yielding $\mathrm{Pos}\ (d_{jo}\ /\ x_1^{k(1)}, \ldots, x_r^{k(r)})$ and $\mathrm{Nec}\ (d_{jo}\ /\ x_1^{k(1)}, \ldots, x_r^{k(r)})$ as great as possible, for instance greater than a given threshold.

Let us denote by S^{jo} the sequence of labels $x_1^{k(1)}, \ldots, x_r^{k(r)}$ for which decision d_{jo} is definitely identified, and by $n_{k(1),\ldots,k(r)}$ the number of times the sequence $x_1^{k(1)}, \ldots, x_r^{k(r)}$ has been obtained as compatible with the opinion expressed by an inquired person in the population. The <u>weighted possibility</u> that a decision d_{jo} of D is satisfying for the considered population can be evaluated as :

$$\Pi(d_{jo}) = \Sigma_{S^{jo}}\ (n_{k(1),\ldots,k(r)}\ /p)\ \ \mathrm{Pos}\ (d_j\ /\ x_1^{k(1)}, \ldots, x_r^{k(r)}).$$

The corresponding <u>weighted necessity</u> equals :

$$N(d_{jo}) = \Sigma_{S^{jo}}\ (n_{k(1),\ldots,k(r)}\ /p)\ \ \mathrm{Nec}\ (d_j\ /\ x_1^{k(1)}, \ldots, x_r^{k(r)}).$$

We consider d_{jo} as <u>satisfying for the whole population</u> if $\Pi(d_{jo})$ corresponds to the highest value and, in case of ambiguity, if $N(d_{jo})$ corresponds to the highest value.

4. CHOICE OF THE SEQUENCES OF QUESTIONS

Let us consider the case where the questionnaire is constructed by successive choices of questions in such a way that we improve the certainty we have about the suitability of a decision of D.

It is obvious that a unique question q_i is necessary if an element d_j of D is identified whatever the answer may be, or, equivalently, if every label q_i^k, for $1 \leq k \leq a(i)$, corresponds to a decision $d_{j(k)}$ such that $\pi(d_{j(k)} / q_i^k) = 1$ and $N(d_{j(k)} / q_i^k) = 1$.

Several questions are generally necessary to come to a decision. The first question to be asked will be q_i, for $1 \leq i \leq m$, processing the most efficient information about the elements of D. We propose to measure this efficiency by means of the <u>certainty provided by question</u> q_i to determine elements of D through its answers is the following :

$$Cer(q_i) = \Sigma_{1 \leq k \leq a(i)} (1 / a(i)) \Sigma_{1 \leq j \leq n} N(d_j / q_i^k),$$

and we choose q_i as a first question if it provides the maximum value for this certainty, among all the possible questions of Q.

Let us suppose that an ordered sequence of questions $x_1, \ldots, x_r$ of Q has been determined. We look for the question which must be chosen in the case where the labels $x_1^{k(1)}, \ldots, x_r^{k(r)}$ have been obtained for every question (see Figure 2).

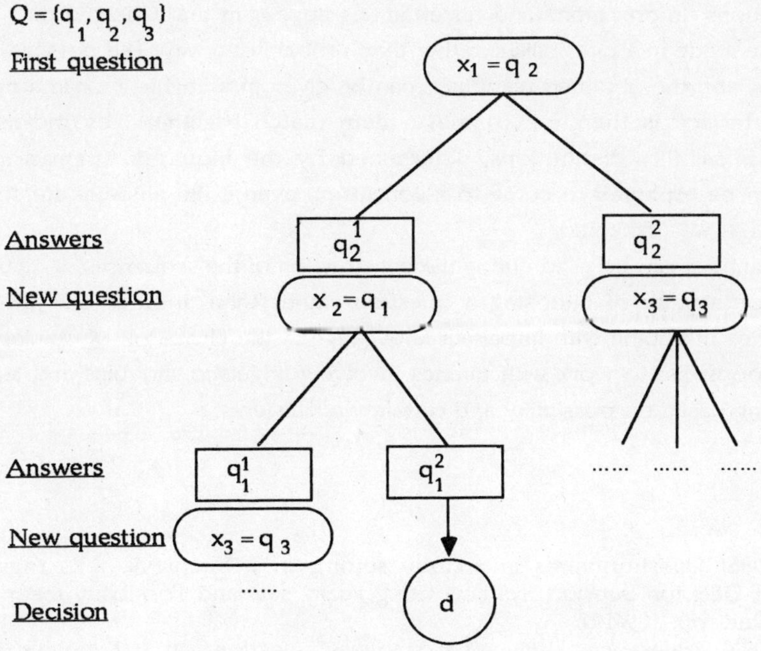

<u>Figure 2 : example of sequences of questions</u>

The certainty provided by the r first questions can be evaluated by :

$$Cer(x_1^{k(1)}, \ldots, x_r^{k(r)}) = \Sigma_{1 \leq j \leq n} Nec(d_j / x_1^{k(1)}, \ldots, x_r^{k(r)}).$$

Let us suppose that we choose question q_i after this sequence of obtained labels. It provides the following increase of certainty with regard to the possible decisions :

$$TCer(q_i) = \Sigma_{1\leq k\leq a(i)}\ Cer(x_1{}^{k(1)}, ..., x_r{}^{k(r)}, q_i{}^k) - Cer(x_1{}^{k(1)}, ..., x_r{}^{k(r)}).$$

Then, we choose the question q_i maximizing this increase of certainty.

The construction of the questionnaire is stopped after a sequence of labels $x_1{}^{k(1)}, ..., x_r{}^{k(r)}$ is obtained for questions $x_1, ..., x_r$, when there exists an element d_{jo} of D such that Pos $(d_{jo} / x_1{}^{k(1)}, ..., x_r{}^{k(r)})$ and Nec $(d_{jo} / x_1{}^{k(1)}, ..., x_r{}^{k(r)})$ are sufficiently high, with respect to a threshold given in a prior way.

5. CONCLUSION

The use of sequences of questions to come to a decision is classical in a probabilistic framework. Nevertheless, it is often difficult for the inquired persons to make a crisp choice for all the questions. Imprecisions and uncertainties appear in the collected data and the decision must be made in a possibilistic rather than probabilistic way. The possible answers to the questions, and the obtained responses can be given in a linguistic or in a numerical way, and an interface is then use to make them match together , by means of the description of possibility distibutions, determined by the inquirers. Imprecisions and uncertainties can be exploited to come to a conclusion even if the answers are somewhat different from what was expected.

Several quantities can be used during the construction of the sequences of questions to evaluate the opportunity of choosing a question rather than another one. This paper proposes a means of coping with imperfect knowledge in the elicitation of opinions and it represents a proposition to work with queries involving linguistic variables and leading to a decision-making based on possibility and certainty coefficients.

REFERENCES

Bouchon B. (1985) Questionnaires in a fuzzy setting, in J. Kacprzyk, R.R. Yager (Eds.), Management Decision Support Systems using Fuzzy sets and Possibility theory, Verlag TUV Rheinland, pp. 189-197.

Bouchon B. (1987) Preferences deduced from fuzzy questions, in J. Kacprzyk and S.A. Orlovski (Eds.), Soft optimization models using fuzzy sets and possibility theory, D. Reidel Publishing Company, pp. 110-120.

Bouchon B. (1988) Questionnaires with fuzzy and probabilistic elements, in J. Kacprzyk, M. Fedrizzi (Eds.), Combining fuzzy imprecision with probabilistic uncertainty in decision making, Springer Verlag, pp. 115-125.

Bouchon B. (1990) Sequences of questions involving linguistic variables, in J.L. Verdegay, M. Delgado (Eds.), Approximate reasoning tools for artificial intelligence, Verlag TUV Rheinland, pp. 115-123.

Dubois D. Prade H. (1987) Théorie des possibilités, applications à la représentation des connaissances en informatique, Masson, Paris.

Farreny H., Prade H., Wyss, E. (1986) Approximate reasoning in a rule-based expert system using possibility theory : a case study, in H.J. Kugler, (Ed.), Information Processing, Elsevier Science Publishers B.V.

Wyss E. (1988) TAIGER, un générateur de systèmes experts adapté au traitement de données incertaines et imprécises, Thèse, Institut National Polytechnique de Toulouse.

SEARCHING FUZZY CONCEPTS IN A NATURAL LANGUAGE DATA BASE

A. F. Rocha * and M. Theoto **

* RANI - Research on Natural and Artificial Inteligence
Rua Tenente Ary Aps, 172 - 13200 Jundiaí - Brazil

** Escola de Enfermagem da USP - Av. Dr. Enéas de C. Aguiar
05403 - São Paulo - Brazil

Abstract: In many instances it is very desirable to store information in a data-base using natural language. As a matter of fact, linguistic variables are key issues in fuzzy data-bases. The complexity of the analysis in retrieving this kind of information may be reduced because, in general, a restricted subset (jargon) of the natural language is used to describe the contents of the data-base. The present paper presents a procedure to recover this jargon. Some examples about how to use this jargon to retrive information from the data-base are presented.

Keywords: Fuzzy Mathematical Programming, Fuzzy Data Base, Natural Language and Data Base, Jargon, Expert Systems.

1. INTRODUCTION

In many instances it is very desirable to store information in a data-base using natural language either to have a user friendly system or because it is the only way to describe the data such as laws, interviews, etc.

At a first glance, one can imagine that the use of natural language could greatly enhance the complexity of analysing and retrieving data from this kind of data-base. However, a careful look into its contents reveals the use of a restricted subset (jargon) of natural language, which greatly reduces the complexity of the semantic analysis required to handle this type of data-base (Sager, 1987; Rocha and Rocha, 1985).

A jargon (J) is composed by a restricted dictionary (D) of terms concerning all words (W) of the language (L), and a small subset (P) of all linguistic productions of this language, using these terms to speak about a reduced subset of meanings (M) of the entire semantic (S) associated to L.

$$J = \{ D, P, M \} \tag{1a}$$

$$P: D \times M \rightarrow [0,1] \tag{1b}$$

$$R: M \times S \rightarrow [0,1] \tag{1c}$$

The restriction (R) upon the semantic (S) is either the expert knowledge in a field of specialization of the human activity (e.g., medicine, law, etc.), or the subject of the inquiring interview, or the scope of the data-base itself (Rocha and Rocha 1985). The knowledge of the jargon is the key issue in rendering manageable a Natural Language Data Base (NLDB) (Sager, 1987).

The first step in characterizing the jargon is to obtain its dictionary D, which components must describe the most significant meanings in the restricted semantic M. Automatization of this process is highly desirable to avoid the necessity of inquiring the expert or the manager of the data-base about the used jargon. Although full automatization is almost impossible with the actual state of art, it is possible to reduce the amount of specific knowledge involved in this task by taking advantage of the relations among semantic and entropy (Shannon, 1974). If this reduction is attainable then it is possible to involve less skilled or naive technicians in the non-automatic steps of obtaining the jargon.

The purpose of the present paper is to describe a procedure that allows a great degree of automatization in the process of obtaining the jargon dictionary by analysing the entropy of the words in a data-base, and using fuzzy mathematical programming (Verdegay, 1984,1989, Delgado, Verdegay and Vila, 1989) for choosing the terms of an initial list (W) of presumptive words. The jargon dictionary (D) is then obtained by deleting irrelevant terms such as articles, conjunctions, etc. from and by adding synonyms to this initial list W. The first task may be automatically performed if the list of irrelevant words is pre-defined, and less skilled or naive technicians may be responsible for the second task. Thus the construction of D requires the optimization of a list of relevant words.

2. OPTIMIZATION OF THE JARGON DICTIONARY

The jargon dictionary (D) must provided the highest amount of information about the contents of the data-base in order to be useful for management of the system.

The amount of information provided by the jargon dictionary is dependent on the entropy of its words (Shannon, 1974, Rocha and Rocha, 1985). Besides this, both very frequent and rare words are meaninless from the management point of view and in most of the instances they are also meaninless for the queries proposed by the users. Therefore, the purpose for obtaining a good dictionary is to maxime its entropy, by eliminating both the most frequent and rare words. These criteria are fuzzy.

The entropy h(D) of the jargon dictionary is calculated from

$$h(D) = - \sum_{i=1}^{n} p_i * \log p_i \qquad (2a)$$

where p_i is the probability of the word $w_i \in D$, and n is the cardinality of D.

Therefore, the maximazation of h(D) may be reduced to the following fuzzy mathematical programming (Delgado, Verdegay and Vila, 1989; Verdegay, 1984):

$$\max h(D) = - \sum_{i=1}^{n} p_i * \log p_i \qquad (2b)$$

such that

$$h(D)/n > \alpha \qquad (2c)$$

$$\sum_{i=1}^{n} p_i = 1 \qquad (2d)$$

The restriction in 2c being that of maintaining the mean entropy of the words in the dictionary D above a fuzzy threshold α. This is the constraint to the cardinality of D to maximize h(D).

The processing is performed by an iterate process of incorporating successive words of smaller entropy until the threshold α is reached. Since, this procedure does not guarantee that all words of the same entropy is incorporated to D when α is reached, then the fuzziness μ(α) of α is calculated as

$$\mu(\alpha) = h(IW_\beta) / (W_\beta) \qquad (2e)$$

where

$$IW_\beta = \{ w_i \in D : h(w_i) = \beta \} \qquad (2f)$$

$$W_\beta = \{ w_i \in DB : h(w_i) = \beta \} \qquad (2g)$$

In other words, the fuzziness is defined by the ratio of the entropy of the words with entropy equal to β incorporated to D and the entropy of the words with the entropy equal to β in the data-base or W.

In this condition, all words of entropy equal to β

are incorporated to D if μ(α) ≥ .5
otherwise they are excluded from D (3)

The key problem in this approach is to determine α. The solution is experimental.

3. PROBING THE BOUNDARIES OF D

7 Medical, 3 Juridical and 3 Interview Data Bases and 5 Data Bases of Recalled Texts associated to the understanding of one original text were used to search the boundary α of their jargon dictionaries (Greco and Rocha, 1987, Machado et al, 1990, Theoto et al, 1987, 1988, 1990). All data were storaged in the corresponding NLDB in its original format. No attention was paid to the correctnes of the sentences either from the semantic or sintactic point of view.

The first step of the analysis to define α was to obtain the list (W) of words in the studied NLDB, and then to order it by frequency of its words and alphabetically.

The probability of each word $w_i \in W$ was obtained as

$$p_i = n_i / N \tag{4a}$$

where n_i is the number of occurrences of the word w_i in the data-base and N is the total number of words in NLDB. The mean entropy per word in NLDB was calculated as

$$h(w) = h(W)/N \tag{4b}$$

The value of h(w) correlated well with the type of NLDB, decreasing from the Law NLDB to the Interview NLDB, such that the data bases could be ordered in the following sequence according with the mean entropy per word

Law, Medical, Text, Interview NLDB

This is in accordance with the fact that a high degree of restriction upon the semantic is very desirable in the case of the juridical jargon in order to avoid multiple interpretation of the laws. On the contrary, is very difficult to impose a high restriction on S in the case of the interviews. As a matter of fact, the amount of restriction in this case is directly correlated with the purpose and guidance exercised during the interview. In the present case, this guidance for the Text NLDB was greater than for the Interview NLDB. Expertise in medicine exerts a strong restriction upon S, but in general smaller than jurisprudence, because variability of the diseases are in general greater than that of the juridic acts.

The second step was to empirically set an initial boundary α by inspecting the meaning of the words in the frequency ordered list of W. The meaningfullness of the words in W regarding the subject of the NLDB was assessed by the manager of the data-base. Thus the initial value of α was very subjective. The considered irrelevant terms were deleted from this list, and synonyms were added consulting the alphabetically ordered list.

4. VALIDATING THE BOUNDARIES

The adequacy of this initial value of α to the purpose of handling the data-base was studied using different approaches depending on the type of the NLDB.

1) Juridical NLDB: D was used to guide a linguistic analysis of the data-base in order to define P. The productions in P were then used on queries to NLDB. The results of the queries were used to assesses the adequacy of α (Rocha et al, in preparation).

2) Medical NLDB: D was compared with a list of significative terms provided by experts and the adequacy of α was derived from the intersection between these different lists (Theoto, Koizumi, Margarido and Rocha, 1990). D was also used to recodify the NLDB and the recodified data-base was used to train a Neural Expert System to diagnose the same disease (Machado et al, 1990, Machado and Rocha, 1990). The success of the expert system validate the adequacy of α.

3) Text NLDB: D was used to recover the structure of the recalled phrases related to the text presented to the volunteers in the study about text understanding (Theoto et al, 1987, Theoto 1990 a,b). The adequacy of α was studied by correlating the different "mean" texts obtained from different groups of volunteers. This analysis showed that understanding (the contents of D) depended on itens as expertise, degree of education, etc (Theoto, 1990 a,b).

4) Interview NLDB: the adequacy of α was evaluated by the amount of information recovered from the NLDB concerning the purpose of the inquiring. The complexity of this task augment as the mean entropy in the NLDB decreased.

Whenever necessary, a new value of α was chosen and the entire analysis repeated.

The final values of α where then compared with the value of h(w) obtained for each NLDB. The results presented in fig. 1 showed the following correlation:

$$\alpha = .7 * h(w) \quad (5)$$

with a correlation coefficient of .83 and significance of $p < .01$.

5. CONCLUSION

The procedure presented here to automatize the generation of the jargon dictionary relevant for the analysis of a NLDB is guaranteed by the results obtained handling different kinds of NLDB with different purposes. The process cannot be fully automatized yet, but requires no specialized knowledge to complement the restriction to impose to the semantic of natural language in order to obtain the set of restricted meanings relevant to the NLDB.

This opens new pathways of investigation on the use of restricted languages for handling data-bases, mainly those concerning with the structure of the productions P used to generate queries and to analyse the data-base contents to answer these queries.

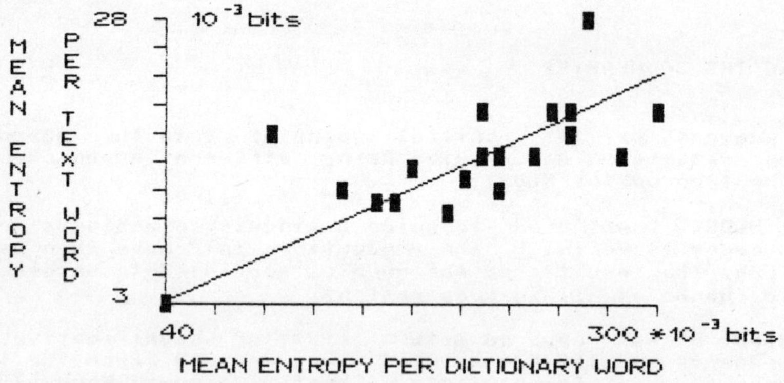

FIG. 1 - DETERMINING THE DICTIONARY BOUNDARY

6. REFERENCES

DELGADO, M., J. L. VERDEGAY AND M. A. VILA - A general model for fuzzy linear programming. Fuzzy Sets and Systems, 1989/29:21-29.

GRECO, G. and A. F. ROCHA - The fuzzy logic of a text understanding. Fuzzy Sets and System, 1987/3:347-360.

MACHADO, R.J.; DUARTE, V.H.A.; DENIS, F.A.R.M. and ROCHA, A.F. - NEXT - The neuro-expert tool. IBM Rio Scientific Center.

MACHADO, R. J. and A. F. ROCHA - The combinatorial neural network: A connectionist model for knowledge based systems. PROCEEDINGS of the 3rd IPMU, Paris, 1990.

ROCHA, A. F. and M. T. ROCHA - Specialized speech: A first prose for expert systems. Information Science, 1985/35: 215-233.

SAGER, N. - Computer processing of narrative information. In Medical Language Processing, Sager, Friedman and Lyman (eds.), Addison-Wesley, 1987:3-22.

SHANNON, E. R. - A mathematical theory of communication. In Key Papers in the Development of the Information Theory, D. Slepien (ed.), IEEE Press, 1974:5-29.

THEOTO, M.; SANTOS, M. R. and UCHIYAMA, N. - The fuzzy decodings of educative texts. Fuzzy Sets and System, 1987/3:331-345.

THEOTO, M. and M. S. KOIZUMI - The expert environment: a case study. PREPRINTS of Second IFSA Congress, vol.1, 1987, Tokyo, 380-383.

THEOTO, M. - Decodification of a Leprosy's Text by students, teachers, nurses and nurses-aid. PhD Thesis (in portuguese). School of Public Health. São Paulo, 1990a.

THEOTO, M. - Text understanding on different populations: A technique for calculation of consensus. PROCEEDINGS of NAFIPS'90, Toronto, 1990b:75-78.

M. THEOTO and M.S. KOIZUMI - Text decoding: an experimental and theoretical approach - PROCEEDINGS of the International Conference on Fuzzy Logic & Neural Networks, Iizuka, Japan, July 20-24, 1990. p.139-142.

THEOTO, M.; KOIZUMI,M.S.; MARGARIDO, L. and A. F. ROCHA - Comparing expertise and the expert's data base. Submitted.

VERDEGAY, J. L. - A dual approach to solve the fuzzy linear programming problem. Fuzzy Sets and Systems, 1984/14: 131-141.

VERDEGAY, J.L. - Fuzzy Mathematical Programming Problem: Resolution. In Systems & Control Encyclopedia, Oxford Singh MG (ed.) Pergamon Press, 1989, 1815-1819.

RECONFIGURABLE NETWORK ARCHITECTURE FOR DISTRIBUTED PROBLEM SOLVING

Jacek Maitan *) and Zbigniew W. Ras **)
*) Lockheed Research Lab., 3251 Hanover Street, Palo Alto, CA 94304-1191, USA
**) University of North Carolina, Dept. of Computer Science, Charlotte, NC 28223, USA

Abstract: OPERA (Opto-Electronic Reconfigurable Architecture) is a new network architecture for multiprocessor systems which features a reconfigurable and adaptable communication structure. This architecture was introduced mainly to suggest a new method of improving the efficiency of solving distributed problems. Distributed problems are decomposed into tasks that have to be dynamically assigned to the nodes of OPERA. Tasks can communicate with other tasks residing at nonlocal nodes through the use of reconfigurable switches. The key objective is to provide a reconfigurable computer architecture for a variety of distributed applications ranging from distributed database systems (Chu, 1982; Maitan, 1989) to multiprocessor clusters at nodes of wide-area networks (Tsai, 1989). The use of tasks which can migrate through the dynamically reconfigurable network in a fashion similar to packetized data introduces new possibilities, especially in the area of fault recovery and load balancing. OPERA may also be useful for solving complex decision making and optimization problems.

Keywords: distributed problem solving, chordal ring network, distributed database, fixed-routing, network backbone, packet-driven routing, reconfigurable network.

1. INTRODUCTION

OPERA (Opto-Electronic Reconfigurable Architecture) is a new network architecture for multiprocessor systems. The key objective is to provide a reconfigurable computer architecture for a variety of distributed applications ranging from distributed database systems (Chu, 1982; Maitan, 1989) to multiprocessor clusters at nodes of wide-area networks (Tsai, 1989). These types of applications require a high level of fault tolerance, in addition to high speed and programmability. The key novel element introduced by OPERA is its scaleability and use of a simple time-sharing mechanism to support a variety of distributed control schemes.

Multigrid Communication Architecture (MCA) (Maitan, 1988), initially designed for wide-area networks, is a foundation for the proposed OPERA structures. These structures can be used to implement backplanes and local area networks (LANs). The key MCA concepts such as:

1) simultaneous use of multiple routing channels,

2) nonlocal routing - a node can skip decoding a data packet from its nearest neighbor, and pass it along a predefined route to another node,

are especially applicable for low-cost, high-performance requirements imposed upon OPERA implementations. These requirements are dictated by the needs of photonics technologies to be used in implementing OPERA's hardware.

In OPERA, all communication must take place in preallocated time slots and use reserved links. These links can be built using fiber-optic technology which allows for many different implementations of separate physical channels. The use of periodic scheduling of preallocated paths (an analog to fast circuit switching), combined with limited buffering in receivers and transmitters, eliminates the need for a real-time routing requiring high-speed decoding, arbitration, and decision-making. As a result, OPERA has the simplest control structure supported by MCA.

2. ARCHITECTURE

Distributed problem solving deals with the issues arising in studies of solving a problem through the interaction of nodes in a network (Singh, 1984). The nodes have a limited expertise, being competent in certain fields of the problem domain. The problem to be solved is decomposed into independent or interrelated subproblems represented by tasks that can be executed in parallel or in a sequential manner. The tasks have to be assigned to the nodes of the network backbone scheme of MCA that interact with each other in order to solve the problem. Each network backbone scheme has many states. Our main goal is to define the set of basic instructions replacing one state of a scheme of MCA by another state. Furthermore, we will suggest a strategy for finding the optimal state of the network backbone for a given problem. This strategy can be identified with a construction of a sequence of basic instructions to be applied one by one to the current state of the scheme in order to build the desired state. The implementation of basic instructions has to be done on the hardware level in terms of switches. In this way, the hardware will be able to reconfigure its structure to the needs presented by software. Following this point of view, our results can be seen as a model of an interface between the hardware and software levels. The network backbone scheme of MCA introduced below is a modification of the definition given in (Gburzynski, 1989).

Definition. By a network backbone scheme of MCA, we mean a sequence $B(X) = < S, P, \Psi, X >$ where:

S is a finite set of stations
P is a finite set of ports
Ψ is a function ($\Psi: P \rightarrow S$) which assigns ports to stations
X is a variable whose values are subsets of $P \times P$ representing sets of links
Ports $p_i, p_j \in P$ belong to the same station if $\Psi(p_i) = \Psi(p_j)$.

By a state of a network backbone scheme B(X) or simply by a network backbone, we mean a sequence $B(L) = < S, P, \Psi, L >$, where $L \subseteq P \times P$.
One state of B(X) will be always distinguished and called an initial state.

Example 1. Let B(L) be a network backbone represented by Figure 1. Each station of B(L) is directly connected to the nearest neighbor and to the station after. All links are point-to-point connections.
We have $S = \{s(i) : 1 \leq i \leq 8\}$, $P = \{p(i) : 1 \leq i \leq 32\}$, and $L = \{l(i) : 0 \leq i \leq 15\}$.

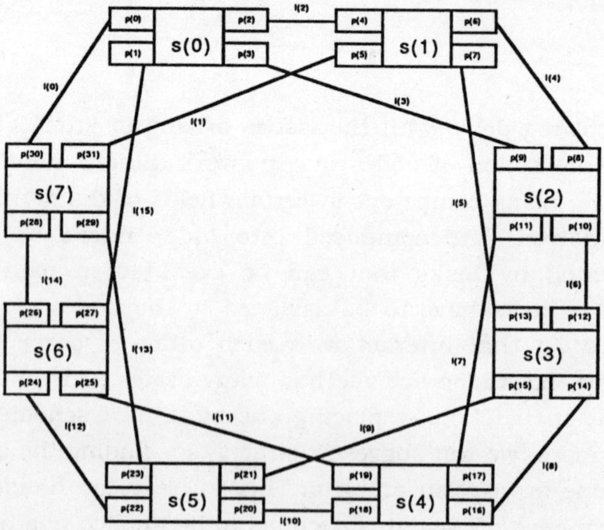

Fig. 1: Chordal ring network with 8 stations.

Function Ψ and the set of links in this particular example are defined by the following equations:

$\Psi(p(i)) = s(\lfloor i/4 \rfloor)$ where $\lfloor i/4 \rfloor$ is defined as the greatest integer which is less than $i/4$, for any i $(1 \leq i \leq 32)$

$l(mod_{16}(2i)) = (\, p(mod_{32}(4i-2)),\ p(mod_{32}(4i))\,)$

$l(mod_{16}(2i+1)) = (\, p(mod_{32}(4i-1)),\ p(mod_{32}(4i+5))\,)$ for any i, $0 \leq i \leq 7$

Let us assume that Figure 1 represents a state of a network backbone scheme B(X). Figure 2 displays the initial state $B(L_0)$ of B(X) and connection patterns between ports which have changed the state $B(L_0)$ to B(L). These new connection patterns (called *fixed-routing*) are represented by bold lines connecting two ports of the same station. They make each station of B(L) directly connected to the station after. The broken lines connecting sets of ports {p(i), p(i+1)} and {p(i+2), p(i+3)} represent the initial connection patterns (called *packet-driven routing*), where $i=4\cdot j$, $1 \leq j \leq 7$. Each change within a station of a network backbone from packet-driven routing to fixed-routing or viceversa is expressed in our terminology as a change from one state of the network to another one.

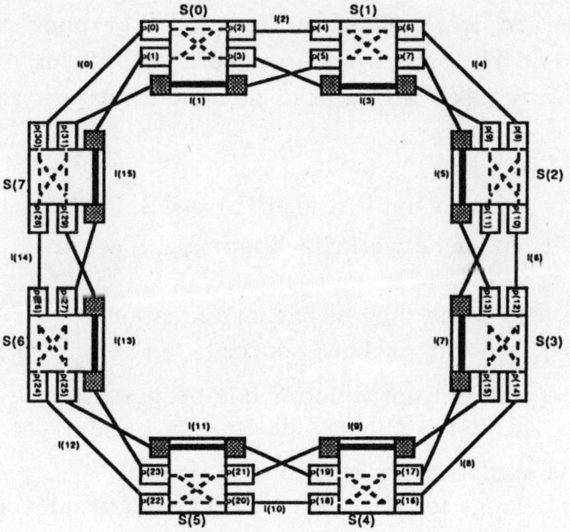

Fig. 2: Chordal ring network with 8 stations (second view).

Let us assume that $B(L_0) = \langle S, P, \Psi, L_0 \rangle$ is an initial state of a network backbone scheme B(X) and $B(L) = \langle S, P, \Psi, L \rangle$.

By L_o-trace for (p,q) /(p,q)∈L/ we mean a shortest sequence of ports $(p_1,p_2,p_3,\ldots,p_{2n-1})$ in B(X) such that there is a sequence of links $(l_1,l_2,l_3,\ldots,l_{n-1})$ in $B(L_o)$ satisfying two conditions below:
$p_1 = p$, $p_{2n-1} = q$ and $(\forall i)(1 \le i \le n-1)[\Psi(p_{2i-1}) = \Psi(p_{2i})$ & $(p_{2i}, p_{2i+1}) = l_i]$.

In our example, the sequence $(p(\mathrm{mod}_{32}(4i-1)), q_1(i), q_2(i), p(\mathrm{mod}_{32}(4i+5)))$, where $\Psi(q_1(i))=\Psi(q_2(i))=s(i)$ represents L_o-trace for $(p(\mathrm{mod}_{32}(4i-1)), p(\mathrm{mod}_{32}(4i+5)))$. The link $l(\mathrm{mod}_{16}(2i+1))$ connects directly ports $p(\mathrm{mod}_{32}(4i-1))$ and $p(\mathrm{mod}_{32}(4i+5))$ through a fixed-routing configuration between ports $q_1(i)$, $q_2(i)$ of the station s(i). It means that stations s(i-1) and s(i+1) can talk directly to each other. Clearly, each fixed-routing configuration through a station reserves its two ports, and the same makes them unavailable for a packet-driven routing. In our example, we had to reserve sixteen ports in order to build the chordal ring network.

3. PROCESS OF DATA TRANSFER

In this section, we discuss the process of transmitting the data between sites of a network backbone. Data are transmitted in the form of packets. They are processed, inserted into, and removed from a network at individual stations. At each station, packets are processed and placed at an output port upon decoded destination address.

Let us assume that DT is a set of data and B(L) = < S, P, Ψ, L> is a network backbone. By a *DT-packet* for a network backbone B(L), we mean a sequence $X=[n,s_1,s_2,d]$ where:

1) $n \in N \cup \{\eth\}$ is the time of packet's creation
2) $s_1 \in S \cup \{\eth\}$ is the packet's source
   (the station from which X has been sent)
3) $s_2 \in S \cup \{\eth\}$ is the packet's destination
   (the station to which X has to be sent)
4) d either belongs to $DT \cup \{\eth\}$ or it is a set/sequence of DT-packets and elements from DT

The packet $\eth=[\eth,\eth,\eth,\eth]$ is called an empty packet. For instance, empty packets can be used for a failure detection in B(L).

The set { $[n_1, s_2, s_{2i}$, ("I have received the following packet",
$[n, s_1, s_3$, "s_1 will be down for 20 minutes"])] : $1 \le i \le 5$}

is an example of the fourth component of a DT-packet. We hope that it clarifies the definition of a DT-packet given above.

Now, we introduce four information retrieval functions, $\sigma, \delta, \tau, \lambda$ used for controlling the routing process in the network backbone B(L). Let $\mathbf{X}(B(L))$ denote the set of all DT-packets for B(L) and $X = [n, s_1, s_2, d] \in \mathbf{X}(B(L))$. Then:

$\sigma : \mathbf{X}(B(L)) \to S$ is the *source address retrieval* function ($\sigma(X) = s_1$)

$\delta : \mathbf{X}(B(L)) \to S$ is the *destination address retrieval* function ($\delta(X) = s_2$)

$\tau : \mathbf{X}(B(L)) \to N$ is the *time stamp retrieval* function ($\tau(X) = n$)

$\lambda : \mathbf{X}(B(L)) \to DATA$ is the *data retrieval* function ($\lambda(X) = d$)

Information storage functions $\sigma^*, \delta^*, \tau^*, \lambda^*$ defined below are used to create packets. If $X = [\_, \_, \_, \_]$ is an arbitrary packet in $\mathbf{X}(B)$, then we have:

$\sigma^*(s, X) = [\_, s, \_, \_]$ where $\sigma^* : S \times \mathbf{X}(B(L)) \to \mathbf{X}(B(L))$

$\delta^*(s, X) = [\_, \_, s, \_]$ where $\delta^* : S \times \mathbf{X}(B(L)) \to \mathbf{X}(B(L))$

$\tau^*(n, X) = [n, \_, \_, \_]$ where $\tau^* : N \times \mathbf{X}(B(L)) \to \mathbf{X}(B(L))$

$\lambda^*(d, X) = [\_, \_, \_, d]$ where $\lambda^* : W \times \mathbf{X}(B(L)) \to \mathbf{X}(B(L))$.

Now, let us denote a set of incoming packets to the station s and a set of outgoing packets from the station s by $X_{in,s}$ and $X_{out,s}$, respectively. The set $X_{in,s}$ is interpreted as a semimemory used to collect all recently received or created packets by the station s. The set $X_{out,s}$ is interpreted as a semimemory used to collect all packets to be considered for distribution to another stations.

Let $B(L) = < S, P, \Psi, L >$ be a network backbone. By a *station* $s \in S$ in B(L), we mean a sequence $s = < \alpha, M, \mathbf{r}_f^*, \mathbf{r}_s, \mathbf{r}_{f,s}, p_f >$ where:

α is the name (address) of the station

M is a memory of the station used for storing packets (it can be interpreted as a set of packets)

$\mathbf{r}_f^* : P \to X_{in,s}$ is a prerouting partial function

$\mathbf{r}_s : X_{in,s} \to P$ is a fixed-routing partial function

$p_f = (p_{f1}, p_{f2})$ is a pair of functions, where:

$p_{f1} : Y_{in,s} \to 2^{\mathbf{X}(B(L))}$ is a preprocessing partial function

where $Y_{in,s} = X_{in,s} - Dom(\mathbf{r}_s)$

$p_{f2} : [\cup p_{f1}(Y_{in,s}) \cup M] \times [\{X_{in,s}\} \times \{M\}] \to X_{out,s} \cup M$

is a processing partial function

$\mathbf{r}_{f,s} : X_{out,s} \to P \cup M \cup X_{in,s}$ is a routing partial function

By $\cup p_{f1}(Y_{in,s})$, we mean $\cup\{ p_{f1}(X) : X \in Y_{in,s}\}$.

Figure 3 shows the design of a single station s. The station has three main components: a memory, a processor, and a switch. The switch is needed for implementing a fixed-routing partial function $\mathbf{r}_s$. The fixed-routing configuration in s is marked by a bold line, whereas packet-driven routing is marked by broken lines. Three ports are reserved for incoming B-packets (read-only ports) and another three for outgoing packets (write-only ports). The station has two semimemories, read-in (X_{in}) and read-out (X_{out}). Both are essential for describing the pair of processing functions (p_{f1}, p_{f2}) and a routing partial function $\mathbf{r}_{f,s}$. For simplicity, we assume that all ports are one-directional. Also, note that information retrieval functions σ, δ, τ, λ can be used for more precise description of routing functions.

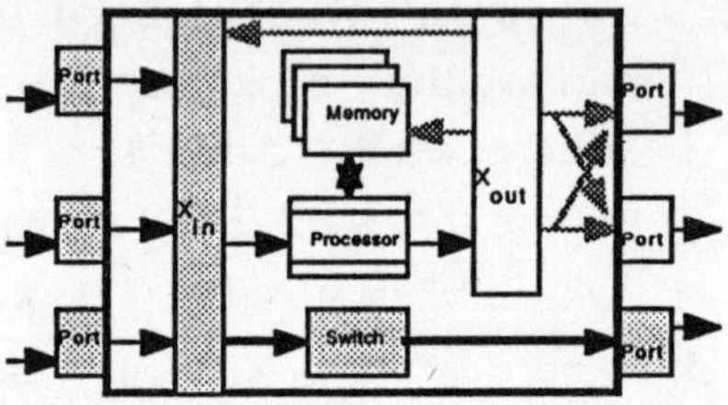

Fig. 3: Example of a station in a network backbone B(L).

Let $B(L) = \langle S, P, \Psi, L \rangle$ be a network backbone and $P_s = \Psi^{-1}(s)$ for any station $s \in S$.

By an *atomic fixed-routing function* in B(L) we mean a binary relation R_{af} on the set of ports P such that:

1. $(p_1, p_2) \in R_{af}$ iff $(\exists s)(p_1, p_2 \in P_s \ \& \ p_1 = \mathbf{r}_s(\mathbf{r}_f^*(p_2)))$
2. R_{af} is one-to-one function

By a *fixed-routing function* in $B(L) = \langle S, P, \Psi, L \rangle$ we mean a binary relation R_f on the set of ports P such that:

$(p, q) \in R_f$ iff there is an L_o-trace $(p_1, p_2, p_3, \ldots, p_{2n-1})$ for (p,q) such that:

$(\forall i)(1 \leq i \leq n-1)[(p_{2i-1}, p_{2i}) \in R_{af}]$

The sequence of links $((p_1, p_2), (p_3, p_4), \ldots, (p_{2n-2}, p_{2n-1}))$ is called a *direct (n-1)-link* connecting ports p, q. The pair (p,q) is its corresponding link in B(L).

Now, we describe the main steps of the process of transferring the data within the network B(L).

Let us assume that we want to send data (denoted by d) from the station s_1 to the station s_2 . The following steps will be performed by the network.

1) The empty packet $\eth=[\eth,\eth,\eth,\eth]$ is created at the station s_1 and it is automatically updated to $X_1=[n,s_1,\eth,\eth]$.

2) $\delta^*(s_2, X_1)$ and $\lambda^*(d, X_1)$ are calculated at the station s_1 and the packet $X_2=[n,s_1,s_2,d]$ is created. This packet is added to the set X_{in,s_1} together with all other packets recently received by ports of s_1. This step is done by the function $\mathbf{r}_f^*$ applied to all ports of the station s_1.

3) If $r_{s_1}(X_2)$ is defined and $\mathbf{r}_{s_1}(X_2) = p$, then X_2 is sent from the station s_1 to another station through the port p (fixed-routing)
else
$p_{f2}((p_{f1}(X_2),\mathbf{X}_{in,s},M_1))$ is computed and the result is sent either to the semimemory X_{out,s_1} or to the memory M_1 of s_1.

4) The routing function $\mathbf{r}_{f,s_1}$ decides which packets from X_{out,s_1} are assigned to which ports for distribution to other stations. Packets not chosen for distribution are sent either to M_1 or X_{in,s_1}.

Clearly, the definitions of $\mathbf{r}_{f,s_1}$ and p_f depend on applications and goals. For example, let us assume that the packet

X = [n, s_1, s_3, "station s_1 will be down for 20 minutes"]

is sent from the station s_1 to the station s_3 through the station s_2 and stations s_{2i} ($i \leq 5$) are neighbors of s_2. Then, the following two examples of preprocessing function p_{f1} associated with the station s_2 are acceptable:

1) $p_{f1}(X)$={[n1,s_2,s_{2i},"s_1 will be down for (20-(n1-n)) minutes"]: $i \leq 5$}

2) $p_{f1}(X)$= {[n1, s_2, s_{2i}, ("I have received the following packet",
[n, s_1, s_3, "s_1 will be down for 20 minutes"])]}.

4. OPTIMIZATION ISSUES

We propose a heuristic algorithm for constructing an optimal network backbone for solving a given problem. So, let $B(X) = < S, P, \Psi, X >$ and P be a problem which has to be solved through the decomposition into subproblems called tasks. These tasks are assigned to stations from $S= \{s_i : i \in I\}$ of the initial network backbone $B(L_o)$ and they need to interact with each other in order to solve P. Let us assume that $(s_{i_1},s_{i_2}),(s_{i_3},s_{i_4}),\ldots,(s_{i_{k-1}},s_{i_k})$ is a sequence of all interactions between stations of $B(L_o)$ needed to solve this problem. Furthermore, let us assume that the list $LIST_1$ is already built. Its elements are defined below:

$$(s_i,s_j,m) \in LIST_1 \quad \text{iff} \quad (s_i,s_j) \in S\times S \text{ and}$$

$$m = card(\{(p,q) \in L_o: p \in \Psi^{-1}(s_i) \,\&\, q \in \Psi^{-1}(s_j)\}) > 0$$

Let N_L: SxS --> N be a function defined as follows:

$$N_L((s_i,s_j))= m \quad \text{iff} \quad (s_i,s_j,m) \in LIST_1.$$

The goal of our heuristic algorithm is to construct an optimal state of B(X) (an optimal network backbone) for solving P. By an optimal state we mean a state in which the total distance between all its stations needed to interact with each other in order to solve P is minimal. We assume that the number of stations which need to talk to each other in order to solve P is not greater than some fixed constant K, for any problem P. The algorithm constructs a list, $LIST_2$, which is used to identify all configurations which should be changed from packet-driven routing to a fixed-routing in order to build an optimal state of B(X) for P. $LIST_2$ is initially empty. It can be easily proved that the time-complexity of this algorithm is $O(n^3 \cdot \log(n))$, where n is the number of stations in $B(L_o)$.

The Algorithm

1. For each $j \in \{1,2...,k-1\}$ and each pair $(s_{i_j},s_{i_{j+1}})$, find the shortest path in $B(L_o)$ linking the coordinates of that pair. We obtain k sequences, where each sequence is of the length less than n-1. Let us denote them by S_i, $1 \le i \le k$.
2. For each subsequence of S_i of the length 3, find the number of occurrences of this subsequence in all S_i, $1 \le i \le k$. Create a decreasing (in terms of the number of occurrences) list LIST of all these subsequences. For instance, $((s_3,s_5,s_8),7)$ would be an example of an element from that list. This element says that (s_3,s_5,s_8) was found 7 times in sequences from S_i, $1 \le i \le k$.
3. Take the first element from LIST. Let us assume that $((s_{i1}, s_{i2}, s_{i3}), m_1)$ is the first element.
4. **If** $[N_L((s_{i1},s_{i2}))>1$ and $N_L((s_{i2},s_{i3}))>1]$ or $\{[N_L((s_{i1},s_{i2}))=1$ implies that s_{i3} always follows (s_{i1},s_{i2}) in all sequences from S_i, $1 \le i \le k]$ and $[N_L((s_{i2},s_{i3}))=1$ implies that s_{i1} always proceeds (s_{i2},s_{i3}) in all sequences from S_i, $1 \le i \le k] \}$

   then assign (s_{i1}, s_{i2}, s_{i3}) to $LIST_2$;
   replace $N_L((s_{i1},s_{i2}))$ by $N_L((s_{i1},s_{i2}))-1$ and
   $N_L((s_{i2},s_{i3}))$ by $N_L((s_{i2},s_{i3}))-1$;
   add $(s_{i1}, s_{i3}, 1)$ to $LIST_1$;
   in each sequence from S_i replace (s_{i1}, s_{i2}, s_{i3}) by (s_{i1}, s_{i3});
   go to 2
   else take the next element from LIST;
   go to 4

The elements of $LIST_2$ have to be interpreted as basic instructions replacing one state of a scheme by another. For instance, a sequence (s_1,s_2,s_3) in $LIST_2$ should be interpreted as a basic instruction defined by the following steps:

1. take a link connecting s_1 with s_2 and denote it by (p_1,p_2)
2. take a link connecting s_2 with s_3 and denote it by (q_2,q_3)
3. connect ports p_2, q_2 of the station s_2 by a fixed routing configuration

Example 2. Let us assume that Fig. 4 represents $B(L_o)$ and (s_1,s_9), (s_1,s_8), (s_5,s_9) is the list of all interactions between stations of $B(L_o)$ needed to solve a problem P. You can verify that $LIST_1$ contains the following elements: $(s_1,s_2,2)$, $(s_1,s_3,1)$, $(s_2,s_3,1)$, $(s_1,s_4,1)$, $(s_2,s_5,1)$, $(s_3,s_4,1)$, $(s_3,s_5,1)$, $(s_4,s_5,1)$, $(s_4,s_6,1)$, $(s_5,s_6,1)$, $(s_6,s_7,1)$, $(s_6,s_8,1)$, $(s_7,s_8,1)$, $(s_7,s_9,2)$, $(s_8,s_9,2)$. Initially, $LIST_2 = \emptyset$.

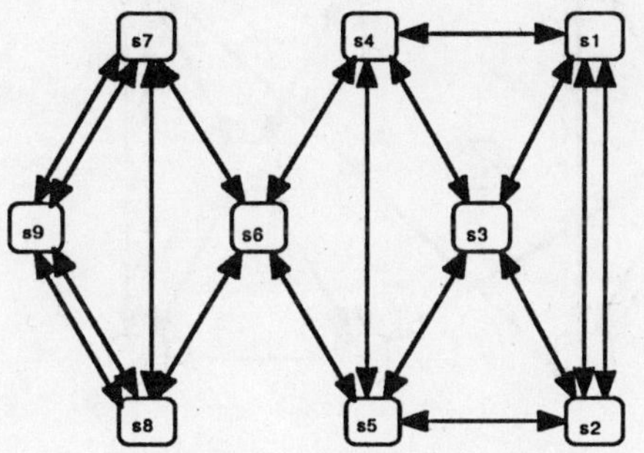

Fig.4: Initial state of a network backbone B(X).

Let us assume that the first step of the algorithm will give sequences:

$S_1 := (s_1, s_4, s_6, s_8)$,

$S_2 = (s_1, s_4, s_6, s_8, s_9)$,

$S_3 = (s_5, s_6, s_7, s_9)$.

The second step will create the list LIST. Its elements in decreasing order are listed below:
$((s_1, s_4, s_6), 2)$, $((s_4, s_6, s_8), 2)$, $((s_6, s_8, s_9), 1)$, $((s_5, s_6, s_7), 1)$, $((s_6, s_7, s_9), 1)$.
The third step takes $((s_1, s_4, s_6), 2)$ from LIST.

Now, $N_L((s_1, s_4))=1$ and s_6 always follows (s_1, s_4) in S_1, S_2 and S_3.

Similarly, $N_L((s_4, s_6))=1$ and s_1 always precedes (s_4, s_6) in S_1, S_2 and S_3.

Hence, (s_1, s_4, s_6) is assigned to $LIST_2$, and $LIST_1$ is updated to:
$(s_1, s_2, 2)$, $(s_1, s_3, 1)$, $(s_2, s_3, 1)$, $(s_1, s_4, 0)$, $(s_2, s_5, 1)$, $(s_3, s_4, 1)$, $(s_3, s_5, 1)$, $(s_4, s_5, 1)$, $(s_4, s_6, 0)$, $(s_5, s_6, 1)$, $(s_6, s_7, 1)$, $(s_6, s_8, 1)$, $(s_7, s_8, 1)$, $(s_7, s_9, 2)$, $(s_8, s_9, 2)$, $(s_1, s_6, 1)$. Sequences S_1, S_2, S_3 are changed to:

$S_1 = (s_1, s_6, s_8)$, $S_2 = (s_1, s_6, s_8, s_9)$, $S_3 = (s_5, s_6, s_7, s_9)$.

The second step will create the new list LIST. Its elements, in decreasing order, are listed below:
$((s_1, s_6, s_8), 2), ((s_6, s_8, s_9), 1), ((s_5, s_6, s_7), 1), ((s_6, s_7, s_9), 1)$.
The third step takes $((s_1, s_6, s_8), 2)$ from LIST. It is true that s_1 always proceeds (s_6, s_8) and s_8 always follows (s_1, s_6) in sequences S_1, S_2, S_3. Hence, (s_1, s_6, s_8) is added to $LIST_2$.
It can be easily checked that in the next 3 loops, (s_5, s_6, s_7) and (s_6, s_7, s_9) will be added to $LIST_2$. The resulting network backbone state is displayed by Fig. 5.

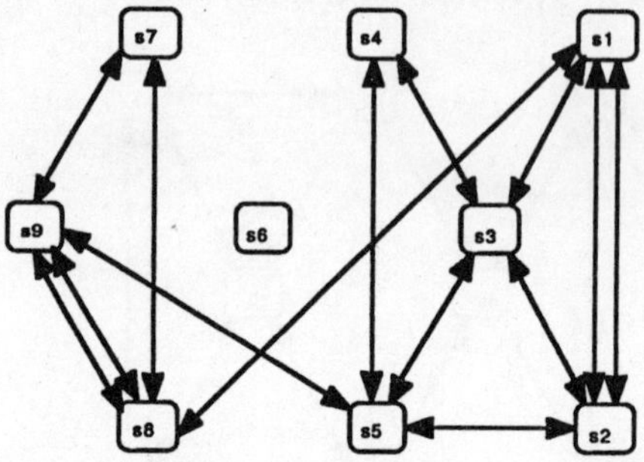

Fig. 5: An optimal state of B(X) for the problem P.

5. CONCLUSION.

Contract net (Davis, 1983) was one of the first concepts addressing the problem of dynamically distributing tasks among processors. A supply-driven strategy for a dynamic distribution of tasks in a multiprocessor system was introduced in (Singh, 1984). In that strategy, all processors place all newly generated tasks, except one task that is chosen as the current task, in the *potential work set*. A processor is allowed to grab a task from the potential work set only when it runs out of internally generated tasks. The problem of a dynamic allocation of tasks for OPERA's processors has a different flavor. Namely, tasks generated by any of the processors can be grabbed only by its neighbors. We have to keep in mind that in OPERA's reconfigurable architecture, the notion of a neighboring processor is dynamic. Two processors can be neighbors in time t and quite far from each other in time t+1. In (Maitan, 1989) we have assumed that the memory of each processor is large enough to store a database and a rule base acquired through learning from other processors. A capability of using both database and rule base suggests new schemes for optimal routing algorithms for database queries within OPERA network architecture. These schemes will be investigated in the context of a hierarchical OPERA network architecture. Our definition of a hierarchical network is based on the use of tree-structured attributes (Michalski, 1983) and differs from definitions based on building hierarchy levels by using recursively distinguished elements from clusters (Tsai, 1989). We hope to be able to demonstrate its advantages against nonhierarchical architectures. We think that the network architecture presented in this paper may be of use in complex decision making and optimization problems too, in particular in the view of recent advances in the computation technology which may better support the solution of those problems.

REFERENCES

Chu, W. (1982), "Optimal query processing for distributed database systems," *IEEE Transactions of Computers*, Vol. C-31, No. 9, pp. 835-850

Davis, R., Smith, R.G. (1983), "Negotiation as a metaphor for distributed problem solving," *Artificial Intelligence*, 20(1)

Gburzynski, P., Rudnicki, P. (1989), "On formal modeling of communication channels," in the *Proceedings of 8th Annual Joint Conf. on Comp. of IEEE Comp. and Comm. Soc.*, pp. 143-151

Maitan, J. (1988), "High-speed routing in fiber-optic networks," Vol. 1, Technical Proposal, LMSC-D089579

Maitan, J., Ras, Z.W., Zemankova, M. (1989), "Query handling and learning in a distributed information system," in Z. W. Ras (ed.), *Methodologies for Intelligent Systems, IV*, North Holland, 118-127

Michalski, R.S., Stepp, R.E. (1983), "Learning from observation: conceptual clustering," in R.S. Michalski, J.G. Carbonell, & T.M. Mitchell (eds.), *Machine Learning: An artificial intelligence approach.* Palo Alto, Tioga Publishing Company

Ras, Z.W. (1990), "Fault-recovery and intelligent distributed systems", in the *Proceedings of ISMVL-90 in Charlotte, N.C.*, IEEE Computer Society, 372-377

Ras, Z.W., Zemankova, M. (1990), "Intelligent query processing in distributed information systems", in Z.W. Ras and M. Zemankova (eds), *Intelligent Systems: State of the art and future directions*, Ellis Horwood Series in Artificial Intelligence, London, England, 357-370

Ras, Z.W. (1991), "Adaptable network architecture for distributed information systems", in *Proceedings of the SPIE's International Symposium on Optical Engineering and Photonics in Aerospace Sensing*, April 1-5, 1991, Orlando, FL

Singh, V., Genesereth, M. (1984), "A variable supply model for distributing deductions," *Technical Report* HPP-84-14, Heuristic Programming Project, Comp. Science Dept., Stanford University

Tsai, W.T., Ramamoorthy, C.V., Tsai, W.K., Nishiguchi, O. (1989), "An adaptive hierarchical routing protocol," *IEEE Transactions on Computers*, Vol. 38, No. 8, pp. 1059-1075

4. ALGORITHMS AND SOFTWARE FOR INTERACTIVE FUZZY OPTIMIZATION

INTERACTIVE DECISION MAKING FOR MULTIOBJECTIVE LINEAR PROGRAMMING PROBLEMS WITH FUZZY PARAMETERS BASED ON A SOLUTION CONCEPT INCORPORATING FUZZY GOALS

MASATOSHI SAKAWA

Department of Industrial and Systems Engineering, Faculty of Engineering, Hiroshima University, Higashi-Hiroshima 724, Japan

HITOSHI YANO

Nagoya Municipal Women's College, Nagoya 464, Japan

ABSTRACT : In this paper, we focus on multiobjective linear programming problems with fuzzy parameters and introduce a new solution concept by assuming that the decision maker may have fuzzy goals for each of the objective functions with fuzzy parameters. By introducing the decision maker's satisficing level for each of the objective functions via the concept of possibility and necessity, we define a new solution concept for multiobjective linear programming problems with fuzzy parameters. Furthermore, we propose an interactive decision making method to derive the satisficing solution of the decision maker from among the proposed solution set. Then the proposed method is illustrated by a numerical example.

KEYWORDS : multiobjective linear programming problems, fuzzy parameters, fuzzy goals, interactive decision making, possibility, necessity.

1. INTRODUCTION

In general, considering the imprecise nature of the human judgements in the real-world decision situations, two types of fuzziness of human judgements should be incorporated in multiobjective programming problems. One is the experts' ambiguous understanding of the nature of the parameters in the problem-formulation process, and the other is the fuzzy goals of the decision maker (DM) for each of the objective functions.

In order to deal with the both types of fuzziness, Sakawa and Yano (1986) formulated the multiobjective linear programming problems with fuzzy parameters which refrect the experts' ambiguous understanding of the nature of the parameters in the problem-formulation process, and proposed interactive fuzzy satisficing method to derive the satisficing solution of the DM by assuming that the fuzzy goals of the DM for each of the objective functions can be quantified by eliciting the corresponding membership functions.

Sakawa *et.al.* (1988, 1989) also defined the four types of α-feasibility and four types of γ-Pareto optimality concepts for multiobjective linear programmming problems with fuzzy parameters by using the four indices proposed by Dubois and Prade (1983) and Dubois (1987) for ranking two fuzzy numbers via the concepts of possibility and necessity. By combining α-feasibility with γ-Pareto optimality, they introduced (α,γ)-Pareto optimal solution for multiobjective linear programming problems with fuzzy parameters.

Buckley (1988) suggested to employ possibility and necessity in multiobjective programming problems. Especially, possiblity distributions are used in place of probability distributions in multiobjective programming problems that involve uncertainty, and possiblity and necessity are both employed to generate solutions. Inuiguchi *et.al.* (1987, 1989) proposed several types of decision procedures via not only possibility and necessity but also impossibility and contingency and explored their utility in multiobjective linear programming problems with fuzzy parameters.

In this paper, by assuming that the fuzzy goals of the decision maker (DM) for each of the objective functions can be quantified by eliciting the corresponding membership functions, we further focus on the multiobjective linear programming problems with fuzzy parameters. To cope with such problems, by adopting two types of indices as the degree of coincidence between the fuzzy goals of the DM and the objective function values involving fuzzy parameters, two types of extended Pareto optimality are defined on the index space. Then a linear programming based algorithm to generate the extended Pareto optimal solutions are developed, and an interactive decision making method to derive the satisficing solution of the DM from among the extended Pareto optimal solution set is proposed.

2. COMPARISON BETWEEN TWO FUZZY NUMBERS

For convenience in our subsequent discussion, first recall the several indices for comparing two fuzzy numbers via the concepts of possibility and necessity. In order to define the inequality relations between two fuzzy numbers $\tilde{m}$ and $\tilde{n}$ as natural extension of ranking two closed intervals, Dubois and Prade (1983, 1987) introduced the following four indices.

Definition 1.

$$Pos(\tilde{m} \geq \tilde{n}) = \sup\{\min(\mu_{\tilde{m}}(u), \mu_{\tilde{n}}(v)) | u \geq v\} \tag{1}$$

$$Pos(\tilde{m} > \tilde{n}) = \sup_{u}\{\inf_{v}\{\min(\mu_{\tilde{m}}(u), 1 - \mu_{\tilde{n}}(v)) | u \leq v\}\} \tag{2}$$

$$Nes(\tilde{m} \geq \tilde{n}) = \inf_{u}\{\sup_{v}\{\max(1 - \mu_{\tilde{m}}(u), \mu_{\tilde{n}}(v)) | u \geq v\}\} \tag{3}$$

$$Nes(\tilde{m} > \tilde{n}) = \sup_{u}\{\max(1 - \mu_{\tilde{m}}(u), 1 - \mu_{\tilde{n}}(v)) | u \leq v\} \tag{4}$$

where $\mu_{\tilde{m}}(u)$ or $\mu_{\tilde{n}}(v)$ is the membership function of $\tilde{m}$ or $\tilde{n}$, and Pos or Nes is short for Possibility or Necessity.

They also defined the following two indices for equality relations between two fuzzy numbers.

Definition 2.

$$Pos(\tilde{m} = \tilde{n}) = \sup_{u}\{\min(\mu_{\tilde{m}}(u), \mu_{\tilde{n}}(u))\} \tag{5}$$

$$Nes(\tilde{m} \supset \tilde{n}) = \inf_{u}\{\max(\mu_{\tilde{m}}(u), 1 - \mu_{\tilde{n}}(u))\} \tag{6}$$

The following two theorems show the relationships between these indices and the α-level sets of the fuzzy numbers, where $m_\alpha^L, m_\alpha^R, n_\alpha^L$ and n_α^R are the left and right side extreme points of the α-level sets of the fuzzy numbers $\tilde{m}$ and $\tilde{n}$ respectively.

Theorem 1.

$$Pos(\tilde{m} \geq \tilde{n}) \geq \alpha \iff m_\alpha^R \geq n_\alpha^L \tag{7}$$

$$Pos(\tilde{m} > \tilde{n}) \geq \alpha \iff m_\alpha^R \geq n_{1-\alpha}^R \tag{8}$$

$$Nes(\tilde{m} \geq \tilde{n}) \geq \alpha \iff m_{1-\alpha}^L \geq n_\alpha^L \tag{9}$$

$$Nes(\tilde{m} > \tilde{n}) \geq \alpha \iff m_{1-\alpha}^L \geq n_{1-\alpha}^R \tag{10}$$

Theorem 2.

$$Pos(\tilde{m} = \tilde{n}) \geq \alpha \iff m_\alpha^R \geq n_\alpha^L \quad \text{and} \quad m_\alpha^L \leq n_\alpha^R \tag{11}$$

$$Nes(\tilde{m} \supset \tilde{n}) \geq \alpha \iff m_\alpha^R \geq n_{1-\alpha}^R \quad \text{and} \quad m_\alpha^L \leq n_{1-\alpha}^L \tag{12}$$

3. PROBLEM FORMULATION

Consider the following multiobjective linear programming problem involving fuzzy parameters (MOLP-FP) (Sakawa and Yano, 1986) :

$$\begin{array}{ll} \min & (\tilde{c}_1 x, \tilde{c}_2 x, \ldots, \tilde{c}_k x) \\ \text{subject to} & x \in X(\tilde{a}, \tilde{b}) = \{x \in R^n \mid \tilde{a}_j x \leq \tilde{b}_j, j = 1, \ldots, m; x \geq 0\}. \end{array} \tag{13}$$

Here $\tilde{c}_i = (\tilde{c}_{i1}, \ldots, \tilde{c}_{in})$, and $\tilde{a}_j = (\tilde{a}_{j1}, \ldots, \tilde{a}_{jn}), \tilde{b}_j$ represent fuzzy parameters refrecting the experts' ambiguous understanding of the nature of the parameters in the problem-formulation process.

We now assume that these fuzzy parameters $\tilde{c}_{i1}, \ldots, \tilde{c}_{in}, \tilde{a}_{j1}, \ldots, \tilde{a}_{jn}$ and $\tilde{b}_j$ in the MOLP-FP are L-R fuzzy numbers (Dubois and Prade, 1980) whose membership functions are continuous and piecewise differentiable, and are denoted by $\mu_{\tilde{c}_{i1}}(c_{i1}), \ldots, \mu_{\tilde{c}_{in}}(c_{in})$, $\mu_{\tilde{a}_{j1}}(a_{j1}), \ldots, \mu_{\tilde{a}_{jn}}(a_{jn})$, and $\mu_{\tilde{b}_j}(b_j)$ respectively.

Now, let $\bar{c}_i = (\bar{c}_{i1}, \ldots, \bar{c}_{in})$, and $\bar{a}_j = (\bar{a}_{j1}, \ldots, \bar{a}_{jn}), \bar{b}_j$ be mean values of the corresponding L-R fuzzy numbers, and consider the following multiobjective linear programming problem (MOLP) where the L-R fuzzy numbers in the MOLP-FP are replaced by the corresponding mean values :

$$\begin{array}{ll} \min & (\bar{c}_1 x, \bar{c}_2 x, \ldots, \bar{c}_k x) \\ \text{subject to} & x \in X(\bar{a}, \bar{b}) = \{x \in R^n \mid \bar{a}_j x \leq \bar{b}_j, j = 1, \ldots, m; x \geq 0\}. \end{array} \tag{14}$$

This problem involves no ambiguity and may be interpreted as the nonfuzzy mathematical model which reflect the greatest possibilities of all of the fuzzy parameters among the MOLP-FP.

In the MOLP, considering the imprecise nature of the DM's judgement, it is natural to assume that the DM may have imprecise or fuzzy goals for each of the objective functions. In a minimization problem, a goal stated by the DM may be to achieve "substantially less than some value A". This type of statement can be quantified by eliciting a corresponding membership function.

When restricting ourselves to a minimization problem, the DM has a fuzzy goal such as "$\bar{c}_i x$ should be substantially less than A_i". In the fuzzy approaches, however, we can further treat a more general case where the DM has two types of fuzzy goals, namely fuzzy goals expressed in words such as "$\bar{c}_i x$ should be in the vicinity of C_i " (called fuzzy equal) as well as "$\bar{c}_i x$ should be substantially less than A_i or greater than B_i " (called fuzzy min or fuzzy max) (Sakawa and Yano, 1986). Such a generalized MOLP (GMOLP) problem may now be expressed as :

$$\begin{array}{rlll} & \text{fuzzy} \quad \min & \bar{c}_i x & (i \in I_1), \\ & \text{fuzzy} \quad \max & \bar{c}_i x & (i \in I_2), \\ & \text{fuzzy} \quad \text{equal} & \bar{c}_i x & (i \in I_3), \\ \text{subject to} & & x \in X(\bar{a}, \bar{b}), & \end{array} \tag{15}$$

where $I_1 \cup I_2 \cup I_3 = \{1, 2, \ldots, k\}, I_i \cap I_j = \phi, \;\; i, j = 1, 2, 3, i \neq j.$

In order to elicit a membership function $\mu_{\tilde{G}_i}(\bar{c}_i x)$ from the DM for a fuzzy goal $\tilde{G}_i$ like "$\bar{c}_i x$ should be in the vicinity of C_i ", it is obvious that we can use different functions to the left and right sides of C_i.

Concerning the membership functions of the DM, it is reasonable to assume that $\mu_{\tilde{G}_i}(\bar{c}_i x)$, $i \in I_1$ and the right side functions of $\mu_{\tilde{G}_i}(\bar{c}_i x), i \in I_3$ are strictly monotone decreasing functions with respect to $\bar{c}_i x$, and $\mu_{\tilde{G}_i}(\bar{c}_i x), i \in I_2$ and the left side functions of $\mu_{\tilde{G}_i}(\bar{c}_i x), i \in I_3$ are strictly monotone increasing functions with respect to $\bar{c}_i x$. To be more explicit, each membership function $\mu_{\tilde{G}_i}(\bar{c}_i x)$ of the DM for $i \in I_1, i \in I_2$ or $i \in I_3$ is defined as follows :

(1) $i \in I_1$:

$$\mu_{\tilde{G}_i}(\bar{c}_i x) = \begin{cases} 1 \quad \text{or} \to 1 & \text{if} \quad (\bar{c}_i x)_R^1 \geq \bar{c}_i x, \\ D_{iR}(\bar{c}_i x) & \text{if} \quad (\bar{c}_i x)_R^1 < \bar{c}_i x < (\bar{c}_i x)_R^0, \\ 0 \quad \text{or} \to 0 & \text{if} \quad \bar{c}_i x \geq (\bar{c}_i x)_R^0, \end{cases} \tag{16}$$

(2) $i \in I_2$:

$$\mu_{\tilde{G}_i}(\bar{c}_i x) = \begin{cases} 0 \quad \text{or} \to 0 & \text{if} \quad (\bar{c}_i x)_L^0 \geq \bar{c}_i x, \\ D_{iL}(\bar{c}_i x) & \text{if} \quad (\bar{c}_i x)_L^0 < \bar{c}_i x < (\bar{c}_i x)_L^1, \\ 1 \quad \text{or} \to 1 & \text{if} \quad \bar{c}_i x \geq (\bar{c}_i x)_L^1, \end{cases} \tag{17}$$

(3) $i \in I_3$:

$$\mu_{\tilde{G}_i}(\bar{c}_i x) = \begin{cases} 0 \quad \text{or} \to 0 & \text{if} \quad \bar{c}_i x \leq (\bar{c}_i x)_L^0, \\ D_{iL}(\bar{c}_i x) & \text{if} \quad (\bar{c}_i x)_L^0 \leq \bar{c}_i x \leq (\bar{c}_i x)_L^1, \\ 1 & \text{if} \quad (\bar{c}_i x)_L^1 \leq \bar{c}_i x \leq (\bar{c}_i x)_R^1, \\ D_{iR}(\bar{c}_i x) & \text{if} \quad (\bar{c}_i x)_R^1 \leq \bar{c}_i x \leq (\bar{c}_i x)_R^0, \\ 0 \quad \text{or} \to 0 & \text{if} \quad (\bar{c}_i x)_R^0 \leq \bar{c}_i x. \end{cases} \tag{18}$$

Here it is assumed that $D_{iR}(\bar{c}_i x)$ or $D_{iL}(\bar{c}_i x)$ is respectively a strictly monotone decreasing or increasing, continuous and piecewise differentiable function with respect to $\bar{c}_i x$ and may be linear or nonlinear, and $(\bar{c}_i x)_L^0$ and $(\bar{c}_i x)_R^0$ are unacceptable levels for $\bar{c}_i x$ and $(\bar{c}_i x)_L^1$ and $(\bar{c}_i x)_R^1$ are totally desirable levels for $\bar{c}_i x$.

After eliciting the membership function $\mu_{\tilde{G}_i}(\cdot)$ of the DM for each of the objective functions in the GMOLP, if we incorporate the fuzzy goals of the DM into the MOLP-FP, the MOLP-FP can be formally transformed as the following fuzzy multiobjective linear decision making problem (FMOLDMP).

$$\begin{array}{ll} \max & (\mu_{\tilde{G}_1}(\tilde{c}_1 x), \mu_{\tilde{G}_2}(\tilde{c}_2 x), \ldots, \mu_{\tilde{G}_k}(\tilde{c}_k x)) \\ \text{subject to} & x \in X(\tilde{a}, \tilde{b}). \end{array} \tag{19}$$

In order to deal with this problem involving fuzzy parameters, we must resolve the following questions.

(1) How can we define the feasibility concept for the constraints involving fuzzy parameters ?
(2) How can we define the degree of the attainment of the fuzzy goal for the objective function value involving fuzzy parameters ?

In order to resolve these two questions, in the following, we introduce new solution concepts for the FMOLDMP by using several indicies for comparing two fuzzy numbers which are introduced in the previous section.

4. α-FEASIBILITY

In order to deal with the constraint set involving fuzzy parameters

$$\tilde{X}(\tilde{a}, \tilde{b}) = \{x \in R^n | \tilde{a}_j x \leq \tilde{b}_j, j = 1, \ldots, m; x \geq 0\}, \tag{20}$$

Sakawa and Yano (1988) have already introduced four types of α-feasibility using the four indices for ranking two fuzzy numbers.

Definition 3. (α-feasibility)

(1) $x \in R^n$ is α-Very Weak feasible (α-VWF) to (20), if and only if

$$x \in X_{VWF}(\alpha) = \{x \in R^n | Pos(\tilde{a}_j x \leq \tilde{b}_j) \geq \alpha, j = 1, \ldots, m; x \geq 0\}. \tag{21}$$

(2) $x \in R^n$ is α-Medium Weak feasible (α-MWF) to (20), if and only if

$$x \in X_{MWF}(\alpha) = \{x \in R^n | Pos(\tilde{a}_j x < \tilde{b}_j) \geq \alpha, j = 1, \ldots, m; x \geq 0\}. \tag{22}$$

(3) $x \in R^n$ is α-Medium Strong feasible (α-MSF) to (20), if and only if

$$x \in X_{MSF}(\alpha) = \{x \in R^n | Nes(\tilde{a}_j x \leq \tilde{b}_j) \geq \alpha, j = 1, \ldots, m; x \geq 0\}. \tag{23}$$

(4) $x \in R^n$ is α-Very Strong feasible (α-VSF) to (20), if and only if

$$x \in X_{VSF}(\alpha) = \{x \in R^n | Nes(\tilde{a}_j x < \tilde{b}_j) \geq \alpha, j = 1, \ldots, m; x \geq 0\}. \tag{24}$$

where the value α $(0 \leq \alpha \leq 1)$ is the threshold for the degree of feasibility.

From the properties of the α-level sets, it can be easily shown that the following theorem holds for $X_{VWF}(\alpha), X_{MWF}(\alpha), X_{MSF}(\alpha)$ and $X_{VSF}(\alpha)$ (Sakawa and Yano 1988).

Theorem 3.

$$X_{VWF}(\alpha) = \{x \in R^n | a^L_{j\alpha} x \leq b^R_{j\alpha}, j = 1, \ldots, m; x \geq 0\}, \tag{25}$$
$$X_{MWF}(\alpha) = \{x \in R^n | a^R_{j,1-\alpha} x \leq b^R_{j\alpha}, j = 1, \ldots, m; x \geq 0\}, \tag{26}$$
$$X_{MSF}(\alpha) = \{x \in R^n | a^L_{j\alpha} x \leq b^L_{j,1-\alpha}, j = 1, \ldots, m; x \geq 0\}, \tag{27}$$
$$X_{VSF}(\alpha) = \{x \in R^n | a^R_{j,1-\alpha} x \leq b^L_{j,1-\alpha}, j = 1, \ldots, m; x \geq 0\}, \tag{28}$$

where $a^L_{j\alpha}$ and $a^R_{j\alpha}$ are n-dimensional row vectors whose ℓ-th elements are $a^L_{j\ell\alpha}$, $a^R_{j\ell\alpha}$, and $a^L_{j\ell\alpha}$, $a^R_{j\ell\alpha}$, $b^L_{j\alpha}$ and $b^R_{j\alpha}$ are the left and right side extreme points of the α-level sets of the fuzzy numbers $\tilde{a}_{j\ell}$ and $\tilde{b}_{j\ell}$ respectively.

It should be noted here that these four types of α-feasibility sets are usual linear constraint sets depending on the degree α $(0 \leq \alpha \leq 1)$.

5. I-PARETO OPTIMALITY

In order to deal with the FMOLDMP, we have to extend the ordinary Pareto optimality concept. To do so, consider the following multiobjective linear programming problems where only the parameters of the objective functions are fuzzy numbers.

$$\begin{array}{ll} \max & (\mu_{\tilde{G}_1}(\tilde{c}_1 x), \mu_{\tilde{G}_2}(\tilde{c}_2 x), \ldots, \mu_{\tilde{G}_k}(\tilde{c}_k x)) \\ \text{subject to} & x \in X = \{x \in R^n \mid a_j x \leq b_j, j = 1, \ldots, m; x \geq 0\}. \end{array} \tag{29}$$

As was pointed out by Inuiguchi *et.al.* (1989), while the fuzziness in the parameters expresses uncertainty of the parameters, the fuzzy goals of the DM for each of the objective functions expresses the indefiniteness in nature or character as idea, feeling, etc. From such a point of view, in this paper, by using the two indices for equality relation between two fuzzy numbers, we adopt the following definition as the degree of the attainment of the fuzzy goals $\tilde{G}_i, i = 1, \ldots, k$.

Definition 4. (satisficing level)

$$\mu_{\tilde{G}_i}(\tilde{c}_i x) \equiv Pos(\tilde{c}_i x = \tilde{G}_i) \tag{30}$$
$$\mu_{\tilde{G}_i}(\tilde{c}_i x) \equiv Nes(\tilde{c}_i x \subset \tilde{G}_i) \tag{31}$$

Observe that the membership function value of $\mu_{\tilde{G}_i}(\tilde{c}_i x)$ is defined as the degree of possibility that the fuzzy goal $\tilde{G}_i$ is equal to the fuzzy objective function $\tilde{c}_i x$, or the degree of necessity that the fuzzy goal $\tilde{G}_i$ includes the fuzzy objective function $\tilde{c}_i x$.

If we apply Definition 4 to the multiobjective linear programming problem involving fuzzy parameters only in the objective functions, the problem can be reduced to the following two types of multiobjective decision making problems respectively.

$$\max_{x \in X} \quad (Pos(\tilde{c}_1 x = \tilde{G}_1), Pos(\tilde{c}_2 x = \tilde{G}_2), \ldots, Pos(\tilde{c}_k x = \tilde{G}_k)) \tag{32}$$

$$\max_{x \in X} \quad (Nes(\tilde{c}_1 x \subset \tilde{G}_1), Nes(\tilde{c}_2 x \subset \tilde{G}_2), \ldots, Nes(\tilde{c}_k x \subset \tilde{G}_k) \tag{33}$$

For these multiobjective decision making problems, as natural extension of usual Pareto optimality, we can define two types of Pareto optimality in the index space as follows.

Definition 5. (I-Pareto optimality)

(1) $x* \in X$ is I-Weak Pareto optimal (I-WP) to (32), if and only if there does not exist another $x \in X$ such that

$$Pos(\tilde{c}_i x = \tilde{G}_i) \geq Pos(\tilde{c}_i x* = \tilde{G}_i), \quad i = 1, \ldots, k. \tag{34}$$

with strictly inequality holding for at least one i.

(2) $x* \in X$ is I-Strong Pareto optimal (I-SP) to (33), if and only if there does not exist another $x \in X$ such that

$$Nes(\tilde{c}_i x \subset \tilde{G}_i) \geq Nes(\tilde{c}_i x* \subset \tilde{G}_i), \quad i = 1, \ldots, k. \tag{35}$$

with strictly inequality holding for at least one i.

Unfortunately, we cannot obtain I-WP or I-SP by directly applying the well-known scalarization methods (Chankong and Haimes, 1983; Sakawa, 1986) for generating Pareto optimal solutions to the usual multiobjective programming problem, since the inequalities (34) and (35) are not yet explicitly represented with respect to x. Therefore, it is necessary to develop some algorithms to generate I-WP to (32) and I-SP to (33).

For that purpose, we assume that some I-WP to (32) or some I-SP to (33) is given, and is denoted by $x* \in X$. For such $x* \in X$, set the parameters β_i or γ_i as follows.

$$\beta_i = Pos(\tilde{c}_i x* = \tilde{G}_i), \qquad i = 1, \ldots, k, \tag{36}$$

$$\gamma_i = Nes(\tilde{c}_i x* \subset \tilde{G}_i), \qquad i = 1, \ldots, k. \tag{37}$$

Then, from Theorem 2, the following theorem directly follows.

Theorem 4.

(1) $Pos(\tilde{c}_i x = \tilde{G}_i) \geq \beta_i, i = 1, \ldots, k$, if and only if

$$c^L_{i\beta_i} x \leq G^R_{i\beta_i}, \qquad i \in I_1 \cup I_3 \tag{38}$$

$$c^R_{i\beta_i} x \geq G^L_{i\beta_i}, \qquad i \in I_2 \cup I_3 \tag{39}$$

(2) $Nes(\tilde{c}_i x \subset \tilde{G}_i) \geq \gamma_i, i = 1, \ldots, k$, if and only if

$$c^R_{i,1-\gamma_i} x \leq G^R_{i\gamma_i}, \qquad i \in I_1 \cup I_3 \tag{40}$$

$$c^L_{i,1-\gamma_i} x \geq G^L_{i\gamma_i}, \qquad i \in I_2 \cup I_3 \tag{41}$$

(Proof)

(1) From (11) in Theorem 2, the following relation holds.

$$Pos(\tilde{c}_i x = \tilde{G}_i) \geq \beta_i, \quad \Longleftrightarrow \quad c^L_{i\beta_i} x \leq G^R_{i\beta_i}, \quad \text{and} \quad c^R_{i\beta_i} x \geq G^L_{i\beta_i}, \quad \text{for any } i \in I_1 \cup I_2 \cup I_3.$$

Since there are three types of the fuzzy goals $\tilde{G}_i$, *i.e.*, fuzzy min ($i \in I_1$), fuzzy max ($i \in I_2$) and fuzzy equal ($i \in I_3$), the above inequalities are equivalent to (38) and (39).

(2) The proof of (2) is similar to the proof of (1) and so omitted.

It should be noted here that (38)-(41) are linear inequalities depending on the degree β_i and $\gamma_i, i = 1, \ldots, k$.

From Definition 5 and Theorem 4, the following theorem is obtained, by which we can check whether $x*$ is I-WP (I-SP) or not.

Theorem 5.

(1) $x*$ is I-WP to (32), if and only if, there does not exist another $x \in X$ such that

$$c^L_{i\beta_i} x \leq G^R_{i\beta_i}, \qquad i \in I_1 \cup I_3 \tag{42}$$

$$c^R_{i\beta_i} x \geq G^L_{i\beta_i}, \qquad i \in I_2 \cup I_3 \tag{43}$$

with strictly inequality holding for at least one i, where β_i is the constant defined by (36).

(2) $x*$ is I-SP to (33), if and only if, there does not exist another $x \in X$ such that

$$c^R_{i,1-\gamma_i} x \leq G^R_{i\gamma_i}, \qquad i \in I_1 \cup I_3 \tag{44}$$

$$c^L_{i,1-\gamma_i} x \geq G^L_{i\gamma_i}, \qquad i \in I_2 \cup I_3 \tag{45}$$

with strictly inequality holding for at least one i, where γ_i is the constant defined by (37).

By using this theorem together with the bisection method, we can now construct the linear programming based algorithm to generate the corresponding I-WP for some given $\beta_i, i = 1, \ldots, k$.

ALGORITHM 1.

Step 1 : Ask the DM to set the degree $\beta_i, (0 \leq \beta_i \leq 1), i = 1, \ldots, k$, in his/her subjective manner.

Step 2 : For the given degree $\beta_i, i = 1, \ldots, k$, using the phase one of the simplex method, check whether an admissible set satisfying the linear inequalities (42) and (43) exists or not.

Step 3 : If an admissible set does not exist, then update β_i to $\beta_i/2, i = 1, \ldots, k$. Otherwise, update β_i to $2\beta_i, i = 1, \ldots, k$. But, if $\beta_{\max} \equiv \max_{i=1,\ldots,k} 2\beta_i > 1$, then update β_i to $2\beta_i/\beta_{\max}, i = 1, \ldots, k$.

Step 4 : If the condition $\Sigma_{i=1}^{k} \mid \beta_i - \beta_i^{'} \mid < \epsilon$ is satisfied, then go to Step 5. Otherwise go to Step 3, where $\epsilon > 0$ is the convergence parameter and $\beta_i^{'}$ is the value of β_i in the last iteration.

Step 5 : For the values of $\beta_i, i = 1, \ldots, k$, solve one of the following linear programming problem depending on $1 \in I_1$ or I_2.

(1) $1 \in I_1$:

$$\begin{array}{lll} \min_{x \in X} & (c_{1\beta_1}^{L} x - G_{1\beta_1}^{R}) & \\ \text{subject to} & c_{i\beta_i}^{L} x \leq G_{i\beta_i}^{R}, & i \in I_1 \cup I_3 \quad i \neq 1 \\ & c_{i\beta_i}^{R} x \geq G_{i\beta_i}^{L}, & i \in I_2 \cup I_3 \end{array} \tag{46}$$

(2) $1 \in I_2$:

$$\begin{array}{lll} \min_{x \in X} & (-c_{1\beta_1}^{R} x + G_{1\beta_1}^{L}) & \\ \text{subject to} & c_{i\beta_i}^{L} x \leq G_{i\beta_i}^{R}, & i \in I_1 \cup I_3 \\ & c_{i\beta_i}^{R} x \geq G_{i\beta_i}^{L}, & i \in I_2 \cup I_3 \quad i \neq 1 \end{array} \tag{47}$$

For the case where $1 \in I_3$, solve the linear programming problem (46) or (47) corresponding to right or left side functions of the fuzzy equal membership function.

It should be noted here that this algorithm always converges for any degree β_i, $(0 \leq \beta_i \leq 1)$, $i = 1, \ldots, k$. The convergence point obtained by applying Algorithm 1 for the initial degree $\beta_i, i = 1, \ldots, k$ is in a sense close to DM's requirement or better than that if the degrees β_i, $i = 1, \ldots, k$ are attainable in the index space. From this point of view, the degrees $\beta_i, i = 1, \ldots, k$ specified by the DM may be called the satisficing levels.

The following theorem shows the relationship between the I-WP to (32) and the satisficing levels $\beta_i, i = 1, \ldots, k$ specified by the DM.

Theorem 6.

Let $x(\epsilon)$ be a convergence vector in the decision space of Algorithm 1 for some satisficing levels $\beta_i, i = 1, \ldots, k$, where $\epsilon > 0$ is the convergence parameter.

(1) If $x(\epsilon)$ is a unique optimal solution of the linear programming problem (46) or (47), then $x* = \lim_{\epsilon \to 0} x(\epsilon)$ is I-WP to (32).

(2) There is some satisficing levels $\beta_i, i = 1, \ldots, k$ such that any I-WP to (32) coincides with the convergence vector $x* = \lim_{\epsilon \to 0} x(\epsilon)$.

The algorithm to obtain I-SP to (33) can also be easily constructed similar to Algorithm 1, and so omitted.

6. I-α-PARETO OPTIMAL SOLUTIONS

Now we are ready to introduce a new solution concept to the FMOLDMP by combining the four types of α-feasibility and the two types of I-Pareto optimality. In the following, we especially focus on the I-α-Very Weak Pareto optimality concept defined as follows.

Definition 6. (I-α-Very Weak Pareto optimal solution)
$x* \in X_{VWF}(\alpha)$ is said to be an I-α-Very Weak Pareto optimal (I-α-VWP) to the FMOLDMP, if and only if there does not exist another $x \in X_{VWF}(\alpha)$ such that

$$Pos(\tilde{c}_i x = \tilde{G}_i) \geq Pos(\tilde{c}_i x* = \tilde{G}_i), \quad i = 1, \ldots, k,$$

with strictly inequality holding for at least one i.

It is easy to obtain I-α-VWP to the FMOLDMP by applying Algorithm 1, since $X_{VWF}(\alpha)$ is a linear inequality set depending on the feasibility degree α.

However, in general, I-α-VWP set to the FMOLDMP consist of an infinite number of points, and some kinds of subjective judgement should be added to the quantitative analyses by the DM. The DM must select his/her satisficing solution from among an I-α-VWP set.

If we introduce a general aggregation function

$$\mu_D\big(Pos(\tilde{c}_1 x = \tilde{G}_1), \ldots, Pos(\tilde{c}_k x = \tilde{G}_k), \alpha\big) \tag{48}$$

where α is the degree of feasibility, a general fuzzy multiobjective decision making problem (GFMDMP) is defined by

$$\begin{array}{ll} \max & \mu_D\big(Pos(\tilde{c}_1 x = \tilde{G}_1), \ldots, Pos(\tilde{c}_k x = \tilde{G}_k), \alpha\big) \\ \text{subject to} & x \in P_{VW}(\alpha), \quad \alpha \in [0, 1]. \end{array} \tag{49}$$

where $P_{VW}(\alpha)$ is the set of I-α-VWPs to (32).

If $\mu_D(\cdot)$ can be explicitly identified, then the GFMDMP reduces to a standard mathematical programming problem. However, it seems to be very difficult to identify $\mu_D(\cdot)$ explicitly and as an alternative, it becomes evident that an interaction with the DM is necessary. In the following, we make the following assumption.

Assumption 1.
$\mu_D(\cdot)$ exists and is known only implicitly to the DM, which means the DM cannot specify the entire form of $\mu_D(\cdot)$, but he/she can provide local information concerning his/her preference. Moreover, it is strictly increasing and continuous with respect to the index value, *i.e.*, $Pos(\tilde{c}_i x = \tilde{G}_i)$ or $Nes(\tilde{c}_i x \subset \tilde{G}_i)$, and the degree α.

7. TRADE-OFF RATES AND INTERACTIVE DECISION MAKING

Now having obtained the I-α-VWP by applying Algorithm 1 for the satisficing levels β_i, $i = 1, \ldots, k$ and the feasibility degree α specified by the DM, the DM must either be satisfied with the current I-α-VWP and the feasibility degree α, or update his/her satisficing levels β_i, $i = 1, \ldots, k$ or the degree α.

In order to help the DM express his/her degree of preference, trade-off information among the index values of the I-α-VWP as well as between the degree α and the index values of the I-α-VWP is very useful. Such a trade-off information is easily obtainable since it is closely related to the simplex multipliers of the linear programming problem (46) or (47).

To derive the trade-off information, define the following Lagrangian function L corresponding to the problem (46)

$$\begin{aligned} L = & (-G^R_{1\beta_1} + c^L_{1\beta_1} x) \\ & + \sum_{i \in I_1 \cup I_3} \Pi_{iR}\{-G^R_{i\beta_i} + c^L_{i\beta_i} x\} + \sum_{i \in I_2 \cup I_3} \Pi_{iL}\{G^L_{i\beta_i} - c^R_{i\beta_i} x\} \\ & + \sum_{j=1}^{k} \Lambda_j (a^L_{j\alpha} x - b^R_{j\alpha}) \end{aligned} \tag{50}$$

where Π_{iL}, Π_{iR} and Λ_j are the simplex multipliers corresponding to the constraints of (46).

Here, we assume that the problem (46) has a unique and nondegenerate optimal solution $x*$ satisfying the following conditions,

(1) $\Pi_{iR} > 0, \quad i \in I_1 \cup I_{3R}, \quad i \neq 1,$

(2) $\Pi_{iL} > 0, \quad i \in I_2 \cup I_{3L},$

where $I_3 = I_{3R} \cup I_{3L}, \quad I_{3R} \cap I_{3L} = \phi.$

Then, by applying the sensitivity theorem (Chankong and Haimes, 1983; Sakawa, 1986) under the assumption that the membership functions of the fuzzy numbers $\tilde{c}_{il}$, $\tilde{a}_{jl}$ and $\tilde{b}_j$ are piecewise differentiable, the following relations hold at the optimal solution $x*$

$$\frac{\partial v}{\partial \beta_i} = \Pi_{iR}\left\{-\frac{\partial G^R_{i\beta_i}}{\partial \beta_i} + \frac{\partial c^L_{i\beta_i} x}{\partial \beta_i}\right\} \quad i \in I_1 \cup I_{3R}, i \neq 1 \tag{51}$$

$$\frac{\partial v}{\partial \beta_i} = \Pi_{iL}\left\{\frac{\partial G^L_{i\beta_i}}{\partial \beta_i} - \frac{\partial c^R_{i\beta_i} x}{\partial \beta_i}\right\} \quad i \in I_2 \cup I_{3L}, \tag{52}$$

$$\frac{\partial v}{\partial \alpha} = \sum_{j=1}^{k} \Lambda_j \left\{\frac{\partial a^L_{j\alpha} x}{\partial \alpha} - \frac{\partial b_{j\alpha}}{\partial \alpha}\right\} \tag{53}$$

where

$$v = -G^R_{1\beta_1} + c^L_{1\beta_1} x. \tag{54}$$

From the assumption that $x*$ is a unique optimal solution to (46), the following relation also follows.

$$v* = -G^R_{1\beta_1} + c^L_{1\beta_1} x* = 0 \tag{55}$$

On the other hand, at the intersection of the membership functions of the fuzzy goals $\tilde{G}_1$ and the fuzzy objective function value $\tilde{c}_1 x*$, the slopes θ_1 and θ_2 of the tangents of the membership functions are expressed as :

$$\tan\theta_1 = -\frac{\partial G^R_{1\beta_1}}{\partial \beta_1}, \tag{56}$$

$$\tan\theta_2 = -\frac{\partial c^L_{1\beta_1}}{\partial \beta_1} x * . \tag{57}$$

Then, from the properties of trigonometric functions, the following relation holds.

$$\frac{\partial \beta_1}{\partial v} = (\tan\theta_1 \tan\theta_2)/(\tan\theta_1 + \tan\theta_2) \tag{58}$$

By partially differentiate the above equations with respect to β_i, $i = 2, \ldots, k$ and α, we obtain the following relations.

$$\frac{\partial \beta_1}{\partial \beta_i} = \left\{\frac{\partial v}{\partial \beta_i}\right\}(\tan\theta_1 \tan\theta_2)/(\tan\theta_1 + \tan\theta_2) \quad i = 2, \ldots, k \tag{59}$$

$$\frac{\partial \beta_1}{\partial \alpha} = \left\{\frac{\partial v}{\partial \alpha}\right\}(\tan\theta_1 \tan\theta_2)/(\tan\theta_1 + \tan\theta_2) \tag{60}$$

It should be noted here that these relations express the trade-off rate formulae between the index values and between the index values and the feasibility degree α.

In the case where $1 \in I_2$, we can also obtain the similar trade-off rate formulae.

From the above discussions, we can now construct the following interactive algorithm to derive the satisficing solution of the DM from among an I-α-VWP set.

ALGORITHM 2.

Step 1 : Ask the DM to select the feasibility degree α and the satisficing levels β_i, $i = 1, \ldots, k$ in his/her subjective manner.

Step 2 : For the feasibility degree α and the satisficing levels β_i, $i = 1, \ldots, k$ specified by the DM, obtain the corresponding I-α-VWP together with the trade-off rates between the index values and the trade-off rate between the index values and the feasibility degree α.

Step 3 : If the DM is satisfied with the current index values and the feasibility degree α, stop. Otherwise, the DM must update his/her satisficing levels β_i, $i = 1, \ldots, k$ and/or α by considering the current values of β_i and α together with the trade-off rate information, and return to Step 2.

8. NUMERICAL EXAMPLE

To clarify the concept of I-α-VWP as well as the proposed algorithm for deriving the satisficing solution of the DM, consider the following two objective linear programming problem involving fuzzy parameters.

$$\begin{array}{ll} \text{fuzzy} \quad \min & x_1 + \tilde{c}_1 x_2 \\ \text{fuzzy} \quad \max & \tilde{c}_2 x_1 + x_2 \\ \text{subject to} & x \in X = \{x \geq 0 \mid x_1 + 2x_2 \leq \tilde{b}\} \end{array}$$

where $\tilde{c}_1, \tilde{c}_2$ and $\tilde{b}$ are fuzzy numbers whose membership functions are defined by :

$$\mu_{\tilde{c}_1}(c_1) = \max(1- \mid c_1 + 2.6 \mid, 0)$$
$$\mu_{\tilde{c}_2}(c_2) = \max(1- \mid c_2 - 1.8 \mid, 0)$$
$$\mu_{\tilde{b}}(b) = \max(1- \mid b - 13 \mid /2, 0)$$

Now, for illustrative purposes, suppose that the interaction with the hypothetical DM establishes the following simple linear membership functions for the fuzzy goals $\tilde{G}_1 and \tilde{G}_2$.

$$\mu_{\tilde{G}_1}(G_1) = \begin{cases} 1 & G_1 \leq -5 \\ 1- \mid G_1 + 5 \mid /5, & -5 < G_1 < 0 \\ 0 & G_1 \geq 0 \end{cases}$$

$$\mu_{\tilde{G}_2}(G_2) = \begin{cases} 0 & G_2 \leq 11 \\ 1- \mid G_2 - 16 \mid /5, & 11 < G_2 < 16 \\ 1 & G_2 \geq 16 \end{cases}$$

Also assume that the hypothetical DM selects the initial values of the degree α to be 0.5, and the initial satisficing levels (β_1, β_2) to be (0.72, 0.96). Then, by applying Algorithm 1, the parameters (β_1, β_2) converge to (0.6, 0.8), and the corresponding linear programming problem in Step 5 of Algorithm 1 is expressed as :

$$\begin{aligned} \min \quad & (x_1 + c^L_{1,0.6} x_2 - G^R_{1,0.6}) \\ \text{subject to } & G^L_{2,0.8} - c^R_{2,0.8} x_1 - x_2 \leq 0 \\ & x_1 + 2x_2 \leq b^R_{0.5}, \quad x_1, x_2 \geq 0 \end{aligned}$$

This problem is equivalent to the following form from the definition of the membership function of $\tilde{G}_1, \tilde{G}_2, \tilde{c}_1, \tilde{c}_2$ and $\tilde{b}$.

$$\begin{aligned} \min \quad & (3 + x_1 - 3x_2) \\ \text{subject to } & 15 - (2x_1 + x_2) \leq 0 \\ & x_1 + 2x_2 \leq 12, \quad x_1, x_2 \geq 0 \end{aligned}$$

This linear programming problem has the unique optimal solution $(x_1*, x_2*) = (6, 3)$, and the corresponding simplex multipliers of the constraints (Π_{2L}, Λ) become (5/3, 7/3), which satisfy the nondegenerate condition. From these values and (51)-(53), (56)-(57) and (59)-(60), the trade-off rate $\partial\beta_1/\partial\beta_2$ becomes -31/16, and $\partial\beta_1/\partial\alpha$ becomes -7/32.

9. CONCLUSION

In this paper, by assuming that the DM has the fuzzy goals for each of the objective functions, four types of α-feasibility and two types of I-Pareto optimality are introduced for multiobjective linear programming problems with fuzzy parameters via the concept of possibility and necessity. By combining α-feasibility and I-Pareto optimality concepts, I-α-Pareto optimal solutions are introduced, and the interactive decision making method to derive the satisficing solution of the DM from among an I-α-Pareto optimal solution set is developed on the basis of linear programming.

However, applications to the real-world problems must be carried out in cooperation with a person actually involved in decision making. From such experiences the proposed method must be revised.

REFERENCES

Buckley J.J. (1988) Possibility and Necessity in Optimization. Fuzzy Sets and Systems 25, 1-13.

Chankong V. and Y.Y. Haimes (1983) Multiobjective Decision Making : Theory and Methodology. North-Holland, New York.

Dubois D. (1987) Linear Programming with Fuzzy Data, in J.C.Bezdek (Eds.): Analysis of Fuzzy Information, Vol.3, CRC Press, pp.241-263.

Dubois D. and H. Prade (1980) Fuzzy Sets and Systems : Theory and Applications, Academic Press.

Dubois D. and H. Prade (1983) Ranking Fuzzy Numbers in the Setting of Possibility Theory, Information Science, 30, 183-224.

Inuiguchi M., H. Ichihashi and H. Tanaka (1987) Decision Procedures Based on Model Concept and its Application to Fuzzy Multiobjective Linear Programming Problem, JORSJ, 30, 449-471 (in Japanese).

Inuiguchi M., H. Ichihashi and H. Tanaka (1989) Possibilistic Linear Programming with Measurable Multi-attribute Value Functions, ORSA Journal of Computing, 1 146-158.

Inuiguchi M. and Y. Kume (1989) Modality Goal Programming Problems, JORSJ, 32, 326-351 (in Japanese)

Inuiguchi M., H. Ichihashi and Y. Kume (1989) A Modalistic Formulation of Fuzzy Mathematical Programming, ISCIE, 2, 69-79 (in Japanese)

Sakawa M. (1986) Optimization in Nonlinear Systems with Single & Multiple Objectives, Morikita (in Japanese).

Sakawa M. (1989) Theory and Applications of Fuzzy Sets, Morikita (in Japanese).

Sakawa M. and H. Yano (1986) An interactive fuzzy satisficing method for multiobjective linear programming problems with fuzzy parameters, in Proc. of the 4-th IFAC/IFORS Symposium, Large Scale Systems : Theory and Applications, Zurich, pp.497-502.

Sakawa M. and H. Yano A (1988) Unified Approach for Multiobjective Linear Programming Problems with Fuzzy Parameters, Trans. of the Institute of Electronics, Information and Communication Engineers, Vol.J71-A, 8, 1569-1575 (in Japanese).

Sakawa M., H. Yano and J. Takahashi (1989) Extended Pareto Optimality Concept for Multiobjective Linear Programming Problems with Fuzzy Parameters and Its Properties, Trans. of the Institute of Electronics, Information and Communication Engineers, Vol.J72-A, 4, 678-684 (in Japanese).

FULP-A PC-SUPPORTED PROCEDURE FOR SOLVING MULTICRITERIA LINEAR PROGRAMMING PROBLEMS WITH FUZZY DATA

Heinrich Rommelfanger
Institute of Statistics and Mathematics
University of Frankfurt, Mertonstraße 17-25
D-6000 Frankfurt am Main, Germany

Abstract:
This paper describes a PC-program for solving (multiobjective) fuzzy linear programming models. It is based on the interactive solution process FULPAL and support the decisionmaker in modelling the membership functions of fuzzy data, specifying the aspiration levels, constructing and solving the surrogate crisp LP-models.

Keywords:
Fuzzy Mathematical Programming,(Fuzzy) Aspiration Levels, Fuzzy Objectives, Interactive Process

1. INTRODUCTION

In real decision situations the decision maker (DM) has often not enough formation for specifying all relevant data by crisp real numbers, as this is required in mathematical optimization models. Although probabilistic approaches fail because there is a qualitatively different kind of imprecision which can be described with inexactness, vagueness and ill-definedness. For coming to grips with this kind of non-stochastic imprecision the fuzzy set theory offers a proper framework.

As real numbers can be modelled by means of fuzzy numbers too, the succeeding fuzzy optimization model (1) is a general form of multimultiobjective linear programming problems with vague data.

$$\left.\begin{array}{l} \tilde{Z}_1(\mathbf{x}) = \tilde{C}_{11}x_1 + \ldots + \tilde{C}_{1n}x_n \\ \vdots \\ \tilde{Z}_K(\mathbf{x}) = \tilde{C}_{K1}x_n + \ldots + \tilde{C}_{Kn}x_n \end{array}\right] \rightarrow \text{Max} \tag{1}$$

subject to

$$\tilde{A}_{i1}x_1 + \ldots + \tilde{A}_{in}x_n \lesssim \tilde{B}_i \;, \quad i = 1,2,\ldots,m_1$$

$$\tilde{A}_{i1}x_1 + \ldots + \tilde{A}_{in}x_n \cong \tilde{B}_i \;, \quad i = m_{1+1},\ldots m$$

$$x_j \geq 0 \;, \quad j = 1,\ldots,n \;.$$

This general form of multicriteria linear programming problems with vague data includes the special cases, that

i. some or all objective functions are deterministic,
ii. there is only one objective (K=1),
iii. some or all constraints are crisp or have the soft form

$$a_{i1}x_1 + a_{i2}x_2 + \cdots + a_{in}x_n \lesssim \tilde{B}_i \quad \text{with} \quad a_{ij} \in \mathbf{R}. \tag{2}$$

A general procedure for solving mathematical programming models of type (1) is the interactive solution process FULPAL (FUzzy Linear Programming based on Aspiration Levels), see Rommelfanger (1988, 1990).
Unfortunately, the application of FULPAL requires many arithmetical steps, because for each objective and for each constraint with fuzzy parameters the decision maker (DM) has to construct a fuzzy objective function in form of a membership function.

Therefore the PC-programm FULP was written to support the DM at
i. input of data
ii. specifying maximal and minimal bounds for the objective functions,
iii. fixing aspiration levels,
iv. constructing fuzzy objectives (membership functions)
v. calculating the surrogate crisp LP-models.

In this paper the software-program FULP and the essential ideas of the underlying solution process FULPAL are described by means of the following numerical example.

$$\left.\begin{array}{llllll} \tilde{Z}_1(\mathbf{x}) = \tilde{C}_{11}x_1 & + \tilde{C}_{12}x_2 & & & + 2x_5 \\ \tilde{Z}_2(\mathbf{x}) = 3x_1 & + \tilde{C}_{22}x_2 & + \tilde{C}_{23}x_3 & + x_4 & + x_5 \\ \tilde{Z}_3(\mathbf{x}) = x_1 & & + 2x_3 & + 5x_4 & \end{array}\right] \rightarrow \text{Max}$$

subject to (4)

$$
\begin{aligned}
\tilde{A}_{11}x_1 + x_2 + \tilde{A}_{13}x_3 + \tilde{A}_{15}x_5 &\lesseqqgtr \tilde{B}_1 \\
3x_1 + \tilde{A}_{22}x_2 + \tilde{A}_{23}x_3 + 2x_4 - 4x_5 &\eqsim \tilde{B}_2 \\
\tilde{A}_{32}x_2 + \tilde{A}_{33}x_3 &\gtrsim \tilde{B}_3 \\
x_1 + 2x_2 + x_3 - x_4 &\lesssim \tilde{B}_4 \\
4x_1 + 2x_4 + x_5 &\le 56 \\
x_1, x_2, x_3, x_4, x_5 &\ge 0
\end{aligned}
$$

2. MODELLING OF THE FUZZY DATA

As the fuzzy data $\tilde{C}_{kj}$, $\tilde{A}_{ij}$, $\tilde{B}_i$ are not yet specified, the DM has to describe them by declaring their membership functions. In practise, all fuzzy data in (1) or in (4) can be modelled by fuzzy numbers or fuzzy intervals of the L-R-type. In numerical examples the authors mostly use the simplest form and take the reference function $R(u) = \mathrm{Max}(0, 1-u)$, i.e. triangular or trapezoid fuzzy sets. Unfortunately the question for the correct form of the membership functions of fuzzy data is generally neglected in literature, although the specifying of membership functions is very important for the acceptance of fuzzy mathematical programming.

In our opinion there exist only few cases, where the DM is capable to describe the precise form of a fuzzy coefficient $\tilde{A}_{ij}$ or $\tilde{C}_{ij}$ or of a soft constraint $\tilde{B}_i$. In general he only knows some salient points.

As a practical way of getting suitable membership functions, we propose that the DM begins with the specification of some prominent membership levels.

For example, we specify the levels $\alpha = 1$, $\alpha = \lambda_A$, $\alpha = \epsilon$ and interprete them for a right side $\tilde{B}_i = (b_i; 0, \bar{\beta}_i)_{RR}$ in a "$\lesssim$"- constraint as follows:

$\alpha = 1$: $\mu_{Bi}(y) = 1$ means that the value y with certainty belongs to the set of available values. The only real number with this attribute is b_i, i.e. $\mu_{Bi}(b_i) = 1$.

$\alpha = \lambda_A$: $\mu_{Bi}(y) \ge \lambda_A$ means that the DM is willing to accept y as an available value for the time being. A value y with $\mu_{Bi}(y) \ge \lambda_A$ has a good chance of belonging to the set of available values.

Corresponding values of y are relevant to the decision. Obviously, a value y with $\mu_{Bi}(y)=\lambda_A$ is a sort of aspiration level.

$\alpha = \epsilon$: $\mu_{Bi}(y) < \epsilon$ means, that y has only a very little chance of belonging to the set of available value The DM is willing to neglect the values y with $\mu_{Bi}(y) < \epsilon$.

Subsequently the DM has to fix values $\bar{b}_i^{\lambda A}$ and $\bar{b}_i^{\epsilon}$ such that $\mu_{Bi}(\bar{b}^{\lambda}A) = \lambda_A$ and $\mu_{Bi}(\bar{b}_i^{\epsilon}) = \epsilon$.

Then the polygon line from $(b_i,1)$ over $(\bar{b}^{\lambda}A, \lambda_A)$ to $(\bar{b}_i^{\epsilon},\epsilon)$ is a suitable approach to μ_{Bi} on the interval $[b_i,\bar{b}_i^{\epsilon}]$.

For all $y \in [b_i,\bar{b}_i^{\epsilon}]$ we set $\mu_{Bi}(y) = 0$, see figure 1.

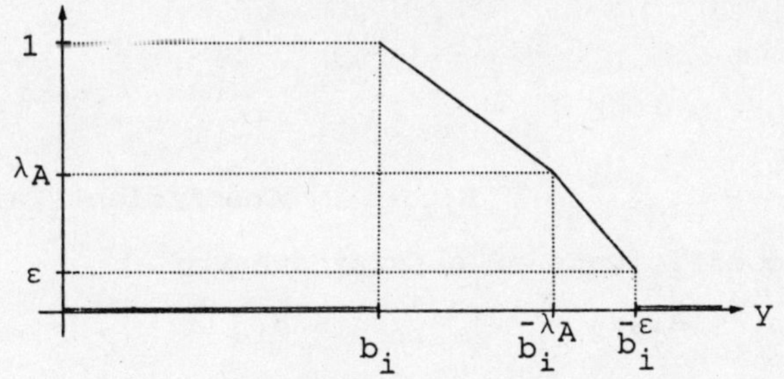

Figure 1: Membership function of $\tilde{B}_i$

Taking pattern from L-R-type fuzzy numbers we symbolize a fuzzy number with this special membership function by

$\tilde{B}_i = (b_i ;0 ,0 ;\bar{\beta}_i^{\lambda A} ,\bar{\beta}_i^{\epsilon})^{\lambda A,\epsilon}$ where

$\bar{\beta}_i^{\lambda A} = \bar{b}_i^{\lambda A} - b_i$ and $\bar{\beta}_i^{\epsilon} = \bar{b}_i^{\epsilon} - b_i$.

If required the DM may specify additional membership levels and additional points $(y,\mu_{Bi}(y))$ on the polygon line.

On the other hand, <u>fuzzy coefficients</u> $\tilde{A}_{ij}$ or $\tilde{C}_{kj}$ do not include elements that may be realized with certainty. In this case, an appropriate point of reference for a fuzzy set $\tilde{A}_{ij}$ is the subset $[\underline{a}_{ij},\bar{a}_{ij}] \subset \mathbb{R}$ consisting of the real numbers with the highest chance of realization. Furthermore, the lower the information of the DM the larger is the length of this interval. Thus, fuzzy coefficients can be modelled as fuzzy intervals $\tilde{A}_{ij} = (\underline{a}_{ij} ;\bar{a}_{ij} ;\underline{\alpha}_{ij} ;\bar{\alpha}_{ij})_{LR}$. This formulation contains the special case, where $\tilde{A}_{ij}$ is a fuzzy number with $\underline{a}_{ij} = \bar{a}_{ij} = a_{ij}$.

But we think it less realistic to assume that all coefficients $\tilde{A}_{ij}$ are fuzzy sets with exactly one mean value a_{ij}, as this was done by Ramik; Rimanek (1985) and Slowinski (1986).

In analogy to the construction of the right sides $\tilde{B}_i$, the coefficients A_{ij} may be modelled as (see Figure 2)

$$\tilde{A}_{ij} = (\underline{a}_{ij}, \bar{a}_{ij}; \underline{\alpha}_{ij}^{\lambda_A}, \underline{\alpha}_{ij}^{\epsilon}; \bar{\alpha}_{ij}^{\lambda_A}, \bar{\alpha}_{ij}^{\epsilon})^{\lambda_A,\epsilon}$$

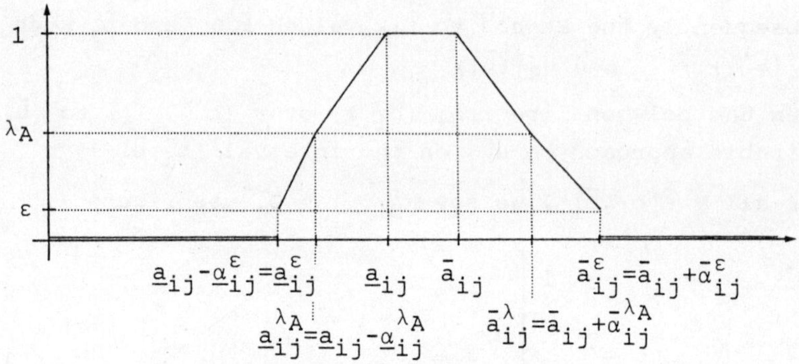

Figure 2: Coefficients $\tilde{A}_{ij}$

If <u>all</u> coefficients of a fuzzy constraint

$$\tilde{A}_{ij}x_1 + \tilde{A}_{i2}x_2 + \ldots + \tilde{A}_{in}x_n \leqq \tilde{B}_i \tag{5}$$

are fuzzy intervals of the above type, the left side of (5) can be summarized with the extension principle to the fuzzy interval

$$\tilde{A}_i(\mathbf{x}) = (\underline{a}_i(\mathbf{x}), \bar{a}_i(\mathbf{x}); \underline{\alpha}_i^{\lambda_A}(\mathbf{x}); \underline{\alpha}_i^{\epsilon}(\mathbf{x});\ \bar{\alpha}_i^{\lambda_A}(\mathbf{x}), \bar{\alpha}_i^{\epsilon}(\mathbf{x}))^{\lambda_A,\epsilon}$$

with $\underline{a}_i(\mathbf{x}) = \sum_{j=1}^{n} \underline{a}_{ij}x_j, \quad \bar{a}_i(\mathbf{x}) = \sum_{j=1}^{n} \bar{a}_{ij}x_j,$

$$\underline{\alpha}_i^{\lambda_A}(\mathbf{x}) = \sum_{j=1}^{n} \underline{\alpha}_{ij}^{\lambda_A}x_j, \quad \bar{\alpha}_i^{\lambda_A}(\mathbf{x}) = \sum_{j=1}^{n} \bar{\alpha}_{ij}^{\lambda_A}x_j \ .$$

$$\underline{\alpha}_i^{\epsilon}(\mathbf{x}) = \sum_{j=1}^{n} \underline{\alpha}_{ij}^{\epsilon}x_j, \quad \bar{\alpha}_i^{\epsilon}(\mathbf{x}) = \sum_{j=1}^{n} \bar{\alpha}_{ij}^{\epsilon}x_i \ .$$

Concerning the numerical example we set $\epsilon = 0.1$. As the spreads of the coefficients $\tilde{A}_{ij}$ and $\tilde{C}_{kj}$ are small in general and considering the inequality relation "$\leqq$"$_R$ we neglect the level λ_A in the case of $\tilde{A}_{ij}$ and $\tilde{C}_{kj}$. The aspiration levels of the right sides $\tilde{B}_i$ will be determined in course of the interactive solution process.

In the numerical example (4) the DM specifies the fuzzy data as follows:

$\tilde{C}_{11} = (4.2\ ,4.5\ ;0.2\ ;0.3)^{0.1}$, $\tilde{C}_{12} = (1\ ,1.3\ ;0.2\ ;0.2)^{0.1}$,
$\tilde{C}_{22} = (4\ ,42\ ;0.3\ ;0.5)^{0.1}$, $\tilde{C}_{23} = (3\ ,3\ ;0.3\ ;0.3)^{0.1}$,
$\tilde{A}_{11} = (3.5\ ,3.8\ ;0.5\ ;0.2)^{0.1}$, $\tilde{A}_{13} = (5\ ,5.6\ ;0.6\ ;0.4)^{0.1}$,
$\tilde{A}_{15} = (1.2\ ,1.5\ ;0.2\ ;0.2)^{0.1}$
$\tilde{A}_{22} = (6.5\ ,7\ ;0.5\ ;0.5)^{0.1}$, $\tilde{A}_{23} = (-3.3\ ,-3\ ;0.2\ ;0.3)^{0.1}$
$\tilde{A}_{32} = (2\ ,2\ ;0.1\ ;0.2)^{0.1}$, $\tilde{A}_{33} = (3.5\ ,3.7\ ;0.2\ ,0.3)^{0.1}$
$\tilde{B}_{1} = (70\ ;0\ ;10)^{0.1}$, $\tilde{B}_{2} = (42\ ,48\ ;5\ ;7)^{0.1}$
$\tilde{B}_{3} = (20\ ;3\ ;0)^{0.1}$, $\tilde{B}_{4} = (40\ ;0\ ;5)^{0.1}$.

3. INEQUALITY REALITIONS

In order to determine a solution to the fuzzy optimization problem (1) or (4) the main question to be answered lies in comparing left and right-hand sides of the constraints

$$\tilde{A}_i(\mathbf{x}) \lesseqq \tilde{B}_i \quad , \tag{6}$$

i.e. the comparison of two (flat) fuzzy numbers.

Obviously it is sufficient to restrict to the "$\lesseqq$"-relation, because a constraint $\tilde{A}_i(\mathbf{x}) \geqq \tilde{B}_i$ is equivalent to $-\tilde{A}_i(\mathbf{x}) \lesseqq -\tilde{B}_i$ and an equation $\tilde{A}_j(\mathbf{x}) \,\tilde{=}\, \tilde{B}_j$ may be replaced by

$$\tilde{A}_j(\mathbf{x}) \lesseqq \tilde{B}_j \quad \text{and} \quad -\tilde{A}_j(\mathbf{x}) \lesseqq -\tilde{B}_j \quad .$$

In literature many interpretations of the inequality relation "$\lesseqq$" in fuzzy constraints are proposed, see the detailed survey in Rommelfanger (1989).

The procedure PC-FULP is based on the interpretation

$$\tilde{A}_i(\mathbf{x}) \lesseqq_R \tilde{B}_i \iff \begin{cases} \sum_{j=1}^{n} (\bar{a}_{ij}+\bar{\alpha}^{\epsilon}_{ij})x_j \le b_i + \bar{\beta}^{\epsilon}_i & \text{(7a)} \\ \mu_{Di}(\sum_{j=1}^{n} \bar{a}_{ij}x_j) \rightarrow \text{Max} & \text{(7b)} \end{cases}$$

It is composed of the "pessimistic" relation (7a), which is used by Slowinski (1986) and other authors too, and the fuzzy goal (7b). The membership function μ_{Di} is defined as

$$\mu_{Di}(y) = \begin{cases} 1 & \text{if} \quad y < b_i \\ \mu_{Bi}(y) & \text{if} \quad y \ge b_i \end{cases} \tag{8}$$

and may be interpreted as the subjective valuation of the needed quantity $\bar{a}_i(\mathbf{x}) = \sum_{j=1}^{n} \bar{a}_{ij}x_j$ with regard to the right-hand side $\tilde{B}_i$.

4. SATISFACTIORY SOLUTIONS AND FUZZY ASPIRATION LEVELS

Obviously,the maximizing of a unique fuzzy objective is a multi-objective problem yet, and this is true more then ever for many fuzzy objectives.

Since a ideal solution of multiobjective problems do not exist in general, we think it better to look for a satisfactory solution.

Let $\tilde{N}_1$ be a fuzzy aspiration level of the goal

$$\tilde{Z}_1(\mathbf{x}) = \tilde{C}_{11}x_1 + \tilde{C}_{12}x_2 + \cdots + \tilde{C}_{1n}x_n \quad \rightarrow \tilde{\text{Max}} \tag{9}$$

In analogy to the description of the right sides $\tilde{B}_i$, the aspiration level $\tilde{N}_1$ can be described by a fuzzy number of the type

$$\tilde{N}_1 = (\underline{n}_1^{\lambda_A}; \underline{\nu}_1^{\epsilon}, \underline{\nu}_1; 0,0)^{\lambda_A,\epsilon} \tag{10}$$

where the right spreads are equal zero.

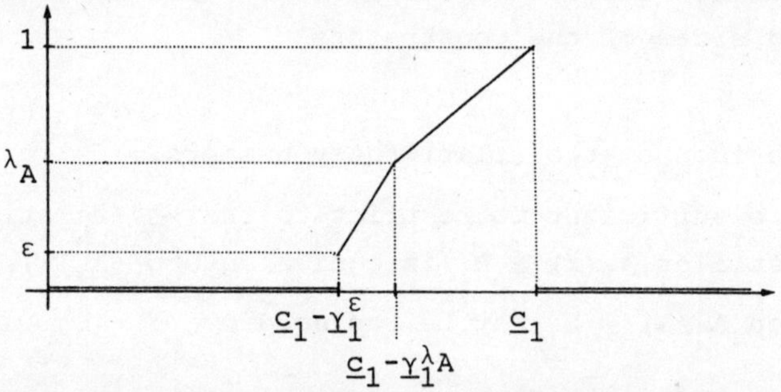

Figure 4: Membership function of μ_{N_1}

When using the inequality relation "$\leqq_R$" the condition

$$\tilde{N}_1 \leqq_R \tilde{Z}_1(\mathbf{x}) \tag{11}$$

is equivalent to

$$\left\{ \begin{array}{l} n_1 - \underline{\nu}_1^{\epsilon} \le \underline{c}_1(\mathbf{x}) - \underline{\gamma}_1^{\epsilon}(\mathbf{x}) = \sum_{j=1}^{n} (\underline{c}_{ij} - \underline{\gamma}_{ij}^{\epsilon}) x_j \qquad (12) \\ \mu_{Z1}(\underline{c}_1(\mathbf{x})) = \mu_{Z1}(\sum_{j=1}^{n} \underline{c}_{1j}x_j) \quad \rightarrow \quad \text{Max} \quad , \qquad (13) \end{array} \right.$$

where μ_{Z1} is the piecewise linear membership function

$$\mu_{Z1}(y) = \begin{cases} 0 & \text{if} \quad y < n_1 - \underline{\nu}_1^{\epsilon} \\ \mu_{N1}(y) & \text{if} \quad n_1 - \underline{\nu}_1^{\epsilon} \le y \le n_1 \\ 1 & \text{if} \quad n_1 < y \quad . \end{cases} \tag{14}$$

For specifying the fuzzy aspiration levels $\tilde{N}_k$ with their characteristic parameters n_k, z_k^A and $n_k - \underline{\nu}_k^{\epsilon}$ it is helpful to calculate the following upper and lower bounds.

Evidently an upper limit for n_k is

$$\bar{z}_k = \underline{c}_k \mathbf{x}_k^{**} = \underset{\mathbf{x}\in X_R}{\mathrm{Max}} \ \underline{c}_k'\mathbf{x} \quad , \tag{15}$$

with $\mathbf{x}' = (x_1,\ldots,x_n)$, $\underline{c}'_k = (\underline{c}_{k1},\ldots,\underline{c}_{kn})$, $k = 1,\ldots,n$

$$X_R = \{\mathbf{x}\in\mathbf{R}_o^n | \sum_{j=1}^{n} (\bar{a}_{ij}+\bar{\alpha}_{ij}^{\epsilon})x_j \le b_i + \bar{\beta}_i^{\epsilon} \quad \text{for any } i = 1,\ldots,n\}$$

On the other hand, $n_k - \underline{\nu}_k^{\epsilon}$ should not be smaller then

$$\underline{z}_k = \mathrm{Min}(\underset{l=1}{\overset{K}{\mathrm{Min}}}(\underline{c}_k-\underline{\gamma}_k^{\epsilon})'\mathbf{x}_l^{*}, \underset{l=1}{\overset{K}{\mathrm{Min}}}(\underline{c}_k-\underline{\gamma}_k^{\epsilon})'\mathbf{x}_l^{**}) \tag{16}$$

where $\mathbf{x}_k^*$ is the solution of the LP-system

$$(\underline{c}_k-\underline{\gamma}_k^{\epsilon})'\mathbf{x} = \sum_{j=1}^{n} (\underline{c}_{kj} - \underline{\gamma}_{kj}^{\epsilon})'x_j \qquad \rightarrow \mathrm{Max}$$

subject to

$$\mathbf{x} \in X_R$$

$$\sum_{j=1}^{n} \bar{a}_{ij}x_j \le b_i \qquad i = 1,\ldots,m$$

The PC-programme FULP calculate these upper and lower bounds automatically and present these to the DM, when he is invited to specify aspiration levels, see the figure 6 in chapter 6.

5. FUZZY OPTIMIZATION

The results in the preceding section can be used directly to transform the fuzzy optimization problem (1) into a deterministic multicriteria mathematical programme.
Employing the inequality relation "$\lessapprox_R$", the FLP (1) can be written as the crisp optimization system

$$(\mu_{Z1}(\underline{c}_1'\mathbf{x}),\ldots,\mu_{ZK}(\underline{c}_K'\mathbf{x}),\mu_{D1}(\bar{\mathbf{a}}_i'\mathbf{x}),\ldots,\mu_{Dm}(\bar{\mathbf{a}}_m'\mathbf{x})) \quad \rightarrow \mathrm{Max}$$

subject to (17)

$$(\underline{c}_k-\underline{\gamma}_k^{\epsilon})'\,\mathbf{x} = \sum_{j=1}^{n} (\underline{c}_{kj}-\underline{\gamma}_{kj}^{\epsilon})x_j \ge n_k - \underline{\nu}_k^{\epsilon} \quad , \quad k = 1,\ldots,K$$

$$(\bar{\mathbf{a}}_i+\bar{\underline{\alpha}}_i^{\epsilon})'\,\mathbf{x} = \sum_{j=1}^{n} (\bar{a}_j+\bar{\alpha}_{ij}^{\epsilon})x_j \le b_i + \bar{\beta}_i^{\epsilon} \quad , \quad i = 1,\ldots,m$$

$$x_j \ge 0 \quad , \quad i = 1,\ldots,m \quad .$$

To get a compromise solution of (17) we use Zadehs min-operator as preference function.
Obviously the compromise objective function

$$\lambda(\mathbf{x}) = \mathrm{Min}(\mu_{Z1}(\underline{c}_i'\mathbf{x}),\ldots,\mu_{Zk}(\underline{c}_K'x),\ \mu_{Di}(\bar{\mathbf{a}}_i'\mathbf{x}),\ldots,\mu_{Dm}(\bar{\mathbf{a}}_{\mathbf{n}}'\mathbf{x}))$$

expresses the total satisfaction of the DM on $\mathbf{x}$.

The inequality relation "$\leqq_R$" has the advantage,that a possible surplus $\bar{a}_i(\mathbf{x})-b_i$ directly influences the decision process.
Moreover "$\leqq_R$" coincides with the usual interpretation of the inequality relation in soft constraints (2), see e.g. Zimmermann (1978), Rommelfanger (1984), Werners (1984) and in the special case of crisp constraints it corresponds to the classical "$\le$"-relation. Thus "$\leqq_R$" is a general definition for inequality relations in optimization models.
When using the inequality relation "$\leqq_R$", it is sufficient for all $\tilde{A}_{ij}$ to specify only the values $\bar{a}_{ij}$ and $a_{ij}^{\epsilon} = \bar{a}_{ij} + \alpha_{ij}^{\epsilon}$.

The PC-programme FULP support the data-input by means of the following algorithm, see figure 3.

```
                    Statement of size of the problem

Please enter the number of objectives
which should be optimized in the problem   :  3

Please enter the number of variables
of the system              :        5

Please enter the number of constraints
of the optimizing model     :       6

Do you want to name the
objectives (e.g. profit, sale) ?  (Y/N)    :  __

                                                            CAPS     Insert
```

```
                    Input of data of the problem

Please specify the coefficients of the objective function.

1. Left (minimal) value of c[3, 5 ] on level alpha = 1      :4.000000
2. Left (minimal) value of c[3, 5 ] on level Epsilon        :3.700000

Please specify the coefficients of the constraints.

1. Right (maximal) value of a[1, 5 ] at level alpha = 1     :1.500000
2. Right (maximal) value of a[1, 5 ] at level Epsilon       :1.700000

Please specify the right sides of the constraints

1. Certain constraint margin b[1 ] on level alpha = 1       :70.00000
2. Maximal constraint margin b[1 ] on level alpha=Epsilon :80.00000

                                                                  NUM
```

Figure 3: Data input

If the set of feasible solutions

$$X_C = \{\mathbf{x}\epsilon X_A \,|\, (\underline{\mathbf{c}}_k - \underline{\gamma}_k^{\epsilon})'\mathbf{x} \geq \underline{n}_k - \underline{\nu}_k^{\epsilon} \quad \text{for any } k = 1,\ldots,K\}$$

is not empty, Negoita and Sularia (1976) have shown that the decision problem

$$\underset{\mathbf{x}\epsilon X_C}{\text{Max Min}} \; (\mu_{Z1}(\underline{\mathbf{c}}_1'\mathbf{x}),\ldots,\mu_{Zk}(\underline{\mathbf{c}}_K'\mathbf{x}),\mu_{D1}(\bar{\mathbf{a}}_i'\mathbf{x}),\ldots,\mu_{Dm}(\bar{\mathbf{a}}_m'\mathbf{x})) \qquad (18)$$

is equivalent to the optimization problem

$$\lambda \rightarrow \text{Max} \qquad (19)$$

subject to

$$k \leq \mu_{Zk}(\underline{\mathbf{c}}_k'\mathbf{x}) \qquad k = 1,\ldots,K$$

$$k \leq \mu_{Di}(\bar{\mathbf{a}}_i'\mathbf{x}) \qquad i = 1,\ldots,m$$

$$\mathbf{x} \in X_C \; ; \qquad \epsilon \leq \lambda \leq 1 \; .$$

Moreover, if all membership functions μ_{Zk} and μ_{Di} are piecewise linear functions, which are concave on $[n_k - \underline{\nu}_k^{\epsilon}, +\infty[$ or $[-\infty, b_i + \bar{\beta}_i^{\epsilon}]$ it has been shown in Rommelfanger (1984) that the system (19) is equivalent to a crisp model. In this special case, the solution of (19) is also a pareto optimal solution of the multicriteria optimization system (17), see Ramik (1987) or Rommelfanger (1988).

In FULPAL, the membership functions f_{Zk} and f_{Di} are constructed as piecewise linear functions with the interpolation nodes

$$\{(n_k - \nu_k^{\epsilon}, \epsilon), (z_k^A = n_k - \underline{\nu}_k^{\lambda}A, \lambda_A), (n_k, 1)\} \text{ or } \{(b_i, 1), (g_i^A = b_i + \bar{\beta}_i^{\lambda}A, \lambda_A), (b_i + \bar{\beta}_i^{\epsilon}, \epsilon)\}$$

and other nodes if necessary.
The PC-programm FULP supports the input of the aspiration levels, see figure 6 and 8.
If one of the membership function μ_{Zk} or μ_{Di}, constructed in this way, is not concave, we can always derive a concave one by a suitable reduction of the ϵ-level set. This correction, which is demonstrated in figure 5, does not influence the solution of the optimization problem because the DM only accepts solutions with $\lambda \geq \lambda_A$.

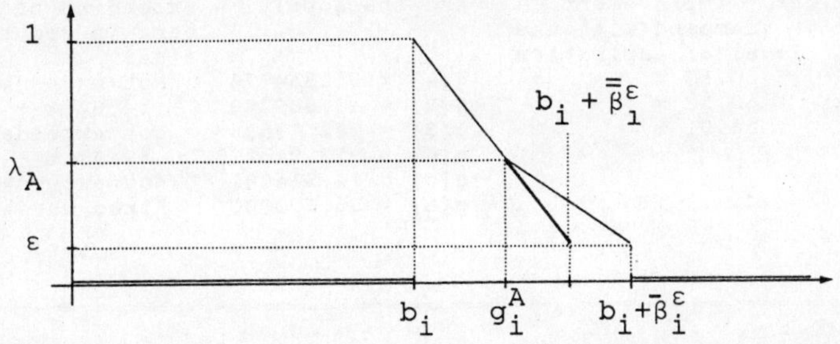

Figure 5: The ϵ-cut of B_i is reduced to $[b_i, b_i + \bar{\bar{\beta}}_i^{\epsilon}]$

This correction is also done by the PC-programme FULP automatically.

6. THE INTERACTIVE SOLUTION PROCESS

The DM specifies the critical level λ_A as λ_A = 0.5 and sets the thresholds for all objectives and all constraints equal 0.1.
For the first iteration of the interactive decision process, the DM intends, to get more information about the solution space. Therefore he begins with the low aspiration levels

$z_1^A = 20 \quad , \quad z_2^A = 70 \quad , \quad z_3^A = 75 \quad ,$

$g_1^A = 79 \quad , \quad g_2^A = 54 \quad , \quad g_3^A = -38 \quad , \quad g_4^A = -17{,}5 \quad , \quad g_5^A = 43$

```
Please specify the levels of aspiration :

concerning the objectives  : z(k)
concerning the constraints : g(i)
-----------------------------------------------------------------
The objective z(3)  must belong to the interval

   [66.6471,102.7974] .

Aspiration level of objective z(3)          =    75


Concerning the constraint g(1)  it must belong to the interval

   [70.0000,80.0000]      .

Aspiration level of constraint g(1)  =    79________________
```

Figure 6: Specify aspiration levels

Using the PC-expert-system "FULP", we get the solution

```
The LP-problem has the following solution :

The critical membership value Lamda-1 = 0.802646

with the following x-values :
x1 = 3.400592    x2 = 10.228536    x3 = 3.321496    x4 = 14.664278
x5 = 13.069078
-----------------------------------------------------------------
the objective values,  improvement          and the supply,   exceeding of the
                       compand with the                       certain right
                    level of aspiration                       sides
z[1] = 50.649176      60.53 %              |g[1] = 61.354776    Not exceeded !
z[2] = 96.676854      60.53 %              |g[2] = 48.889283      14.82 %
z[3] = 97.224338      83.01 %              |g[3] = -42.778567   Not exceeded !
                                           |g[4] = -19.013228     39.47 %
                                           |g[5] = 12.514881    Not exceeded !
                                           |g[6] = 56.000000    Fixed variable.

Press F1 for help
Do you want to print this solution by printer ? (Y/N) __       CAPS      Insert
```

Figure 7: First compromise solution

As $\lambda^1 = 0{,}902645 > \lambda_A$, the levels of aspiration can be increased. The DM decides to change all levels and he specifies

$$z_1^A = 40 \quad , \quad z_2^A = 100 \quad , \quad z_3^A = 90$$

$$g_1^A = 75 \quad , \quad g_2^A = 53 \quad , \quad g_3^A = -39 \quad , \quad g_4^A = -18 \quad , \quad g_5^A = 43$$

In doing so the DM gets helpful assistance by the PC-expert-system FULP, see figure 8.

```
Please specify the new levels of aspiration:

concerning the objectives  z(k)
concerning the constraints g(i)
===========================================================================
Concerning the objective z(2)  the level of aspiration must belong to the

interval  [100.0000,114.0728]  .

Aspiration level of objective  z(2)          =     103.000000________

          z-Min[k]            zA[k]              z[k]              z-Max[k]
            |-------------------|------------------|------------------|
        60.2718                 |                  |              114.0728
                           100.0000                |
                                               101.6335
```

Press function key F1 for help about your input

Figure 8: Specification of the new aspiration level z_2^A

Calculating the new LP-model (LP.2), we get the result:

```
The LP-problem has the following solution :

The critical membership value Lamda-2 = 0.558037

with the following X-values :
x1 = 1.748727    x2 = 10.227904    x3 = 3.322681    x4 = 17.803615
x5 = 13.397863
===========================================================================
the objective values,  improvement          and the supply,   exceeding of the
          |          compand with the                |        certain right
          |        level of aspiration               |        sides
z[1] = 44.368281       14.26 %            |g[1] = 55.576870     Not exceeded !
z[2] = 101.633476      11.61 %            |g[2] = 48.889244       17.78 %
z[3] = 100.352102      80.89 %            |g[3] = -42.778488    Not exceeded !
                                          |g[4] = -18.687326      65.63 %
                                          |g[5] = 7.723599      Not exceeded !
                                          |g[6] = 56.000000     Fixed variable.
```

Figure 9: Second compromise solution

As $\lambda^2 = 0.558037$ indicates that there is only little margin for further improvements, the DM keeps constant the levels z_1^A, z_3^A, $g_2^A, \ldots, g_5^A$. He merely changes the two remainung levels to

$z_1^A = 104$ and $g_1^A = 70$.

The new LP-model (LP.3) has the solution:

```
The LP-problem has the following solution :

The critical membership value Lamda- 3 = 0.512605

with the following x-values :
x1 = 0.888939    x2 = 10.202616    x3 = 3.370095    x4 = 19.468979
x5 = 13.506284

the objective values,  improvement            and the supply,     exceeding of the
            |        compand with the                   |         certain right
            |      level of aspiration                  |         sides
z[1] = 40.948730        3.10 %             |g[1] = 52.712545     Not exceeded !
z[2] = 104.253931       2.52 %             |g[2] = 48.887663       17.75 %
z[3] = 101.554382      90.29 %             |g[3] = -42.775327    Not exceeded !
                                           |g[4] = -18.694281     65.29 %
                                           |g[5] = 5.195287      Not exceeded !
                                           |g[6] = 56.000000     Fixed variable.
```

Figure 10: Third compromise solution

Now the DM accepts this solution.

7. REFERENCES

Delgado M. und J.L.Verdegay and M.A.Vila (1989) A General Model for Fuzzy Linear Programming. FSS 29, 21-30.

Kacprzyk J. und R.R.Yager,Eds.(1985) Management Decision Support Systems using Fuzzy Sets and Possibility Theory. Verlag TÜV Rheinland, Köln.

Luhandjula M.K. (1989) Fuzzy Optimization: An Appraisal. FSS 30, 257-282.

Negoita C.V.and M.Sularia (1976) On Fuzzy Mathematical Programming and Tolerances in Planning. Economic Computation and Economic Cybernetics Studies and Research 3, 3-15

Ramík J.and J.Rímánek (1985) Inequality between Fuzzy Numbers and its use in Fuzzy Optimization. FSS 16, 123-138.

Rommelfanger H. (1984) Concave Membership Functions and their Application in Fuzzy Mathematical Programming. Proceedings of the Workshop on the Membership Function, Ed. by the European Institute for Advanced Studies in Management (EIASM), pp.88-101.

Rommelfanger H. (1988) Entscheiden bei Unschärfe. Fuzzy Decision Support-Systeme. Springer Verlag, Berlin Heidelberg New York London Paris Tokyo.

Rommelfanger H. (1989) Inequality Relations in Fuzzy Constraints and its use in Linear Fuzzy Optimization. in: J.L.Verdegay and M.Delgado (Eds.): The Interface between Artificial Intelligence and Operational Research in Fuzzy Environment. Verlag TÜV Rheinland Köln 1989, pp.195-211

Slowinski R. (1986) A Multicriteria Fuzzy Linear Programming Method for Water Supply System Development Planning, FSS 19, 217-237.

Tanaka H. and K.Asai (1984) Fuzzy Linear Programming with Fuzzy Numbers, FSS 13, 1-10.

Werners B. (1984) Interaktive Entscheidungsunterstützung durch ein flexibles mathematisches Programmierungssystem, Minerva Publikation, München.

Zimmermann H.J. (1978) Fuzzy Programming and Linear Programming with several Objective Functions, FSS 1, 45-55.

'FLIP' : MULTIOBJECTIVE FUZZY LINEAR PROGRAMMING SOFTWARE WITH GRAPHICAL FACILITIES

Piotr CZYŻAK and Roman SŁOWIŃSKI
Institute of Computing Science
Technical University of Poznań
60-965 Poznań, Poland

Abstract: The paper presents a microcomputer implementation of a visual interactive method, called 'FLIP', for solving MOLP problems with fuzzy coefficients in the objective functions and on the both sides of the constraints. The idea of 'FLIP' relies on an observation that for given values of decision variables, the main question to be answered consists in the comparison of objectives and goals, and left- and right-hand-sides of the constraints which are fuzzy numbers. In result of application of a comparison principle, the fuzzy MOLP problem is transformed into a non-fuzzy multiobjective linear fractional programming (MOLFP) problem which is solved using an interactive sampling method. An evaluation of the quality of successive proposals (solutions) is based on the following characteristics: (i) scores of fuzzy objectives in relation to the goals, (ii) dispersion of values of the fuzzy objectives due to uncertainty, (iii) safety of solutions or risk of violation of the constraints. The graphics displayed by 'FLIP' provide the most comprehensive synthesis of these characteristics. The aim of this paper is to demonstrate that the kind of visual interaction organized by 'FLIP' is the most appropriate for solving MOLP problems under uncertainty and imprecision. In order to support this conclusion, a land-use planning problem is solved using the 'FLIP' software on a micro.

Keywords: Fuzzy Linear Programming, Multiple Objectives, Interactive, Graphical Display of Fuzzines, Software

1. INTRODUCTION

In recent years, we have been able to observe a growing interest in multiobjective linear programming (MOLP) under uncertainty and imprecision. This follows from the fact that deterministic multiobjective optimization models are often insufficient in practical situations, especially in long-term planning problems, development strategies and agricultural decision problems. The coefficients that appear in MOLP problems may not be well-defined, either because their value depends upon other parameters (not accounted for in the model) or because they cannot be precisely assessed, and only qualitative estimates of these coefficients are available. Moreover, the constraints expressed by equalities or inequalities between linear expressions are often softer in reality that what their mathematical expression might let us believe.

This situation has motivated a search for more flexible formulations of optimization models that, although remaining rigorous, may help bridging the gap between the mathematical models and the real decision-making situations, through handling of uncertainty and imprecision. Two distinct lines of research try to address these issues: stochastic linear programming and fuzzy linear programming that have developed independently. The first comparative study of stochastic and fuzzy approaches to MOLP has been performed by Słowiński and Teghem (1988). Recently, a more comprehensive analysis of both approaches has been described in a multi-author book edited by Słowiński and Teghem (1990). The book is composed of four parts. The first part contains three tutorial chapters related to three key components of the study: multiobjective, stochastic and fuzzy programming, separately. The second and the third part group chapters representing stochastic and fuzzy methodologies, respectively, while the last part collects chapters in which stochastic and fuzzy approaches are considered together. Being written by leading specialists of each approach, this is the most complete source book on multiobjective mathematical programming under uncertainty and imprecision.

In this paper, we are dealing with a method, called 'FLIP', representing the fuzzy approach to modelling of uncertainty and imprecision. Its theoretical foundations have been described first in Słowiński (1986) and then developed in Słowiński (1990). It has been applied for solving many real-life decision problems, in particular, water supply planning problem (Słowiński, 1986; Słowiński, Urbaniak, Węglarz, 1986; Słowiński, 1987), diet optimization for farm animals

(Czyżak, 1989; Czyżak, Słowiński, 1989; Czyżak, Słowiński, 1990) and farm structure optimization (Czyżak, 1990). In all these applictions, the experts taking part in model building have accepted to express uncertain or imprecise parameters in terms of tolerance intervals with a most possible value (or subinterval) and decreasing possibility for other values within the interval. This corresponds exactly to the definition of fuzzy numbers, i.e. normal convex continuous fuzzy subsets of the real line. So, the modelling of uncertainty and imprecision using fuzzy numbers was quite natural there.

The aim of the present paper is to characterize the most recent microcomputer implementation of 'FLIP' and to present its application to land-use planning problem. The idea of 'FLIP' will be recalled in the next section. The interactive process organized by 'FLIP' will be described in section 3. In the last section, the land-use planning problem will be solved using the 'FLIP' software on a micro.

2. FOUNDATIONS OF THE 'FLIP' METHOD

'FLIP' solves the following MOLP problem with fuzzy coefficients:

$$\widetilde{\text{Maximize}} \begin{bmatrix} \tilde{z}_1 = \underline{\tilde{c}}_1 \underline{x} \\ \dots\dots\dots \\ \tilde{z}_k = \underline{\tilde{c}}_k \underline{x} \end{bmatrix} \tag{1}$$

$$\text{s.t.} \quad \underline{\tilde{a}}_i \underline{x} \lesssim \tilde{b}_i \qquad i = 1,\dots,m \tag{2}$$

$$\underline{x} \geq \underline{0} \tag{3}$$

where $\underline{x}$ is a column vector of n decision variables, $\underline{\tilde{c}}_1,\dots,\underline{\tilde{c}}_k$ are row vectors of fuzzy cost coefficients corresponding to the objective functions $\tilde{z}_1,\dots,\tilde{z}_k$, $\underline{\tilde{a}}_i$ is the i-th row of the matrix of fuzzy coefficients, and $\tilde{b}_i$ is its corresponding fuzzy right-hand-side. Notice that a "greater then or equal to" constraint can be transformed to (2) by multiplying the constraint by -1, and equality constraint can be presented as a pair of weak inequality constraints with opposite relation signs. It is assumed, moreover, that for the particular objectives, the decision maker (DM) is in a position to define fuzzy aspiration levels, thought of as goals, denoted by $\tilde{g}_1,\dots,\tilde{g}_k$.

All fuzzy coefficients are given as L-R type fuzzy numbers, i.e. number $\tilde{a}$ is a triple of parameters (a,α,β) of its membership function

$$\mu_a(x)=\begin{cases} L((a-x)/\alpha) & \text{if } x \leqslant a \\ R((x-a)/\beta) & \text{if } x \geqslant a \end{cases}$$

where a is the "most possible" value, α and β are nonnegative left and right "spreads" of $\tilde{a}$, respectively, and L, R are symmetric bell-shaped reference functions that are decreasing in $(-\infty\ ,\ \infty)$ and $L(0)=R(0)=1$, $L(1)=R(1)=0$. When the spreads are zero, then $\tilde{a}$ is a nonfuzzy (crisp) number equal to a.

The idea of FLIP relies on an observation that for a given $\underline{x}$, the main question to be answered consists in the comparison of the left- and right-hand-sides in objectives and constraints which are fuzzy numbers. Assuming an L-R representation of fuzzy coefficients, a comparison principle has been proposed (Słowiński, 1986, 1990) which allows a transformation of the fuzzy MOLP problem into a multiobjective linear fractional programming problem. The best compromise solution of the latter problem ensures the "best consistency" between the goals and the objective functions, and satisfies the constraints with a given "safety".

Let us recall informally the comparison principle using an example of constraint i for a given $\underline{x}$. Fuzzy coefficients of the constraints are given as L-R fuzzy numbers:

$$\tilde{\underline{a}}_i = (\underline{a}_i,\underline{\alpha}_i,\underline{\beta}_i)_{LR}\ ,\quad \tilde{b}_i = (b_i,\gamma_i,\delta_i)_{RL}\ ,\qquad i=1,...m.$$

For the sake of clarity, we assume that the reference functions of all fuzzy coefficients are of two kinds only: L and R. This is not, however, a general assumpion of the comparison principle (cf. Słowiński, 1990).

It should be specified that all arithmetic operations on fuzzy numbers taking place in (1)-(3) are extended operations in the sense of Zadeh's principle. For any $\underline{x} \geq \underline{0}$, the left-hand-side of the i-th constraint can be summarized to the fuzzy number:

$$\tilde{\underline{a}}_i\underline{x} = (\underline{a}_i\underline{x},\ \underline{\alpha}_i\underline{x},\ \underline{\beta}_i\underline{x})_{LR}\ ,\qquad i=1,...,m.$$

Fig.1 shows relative positions of fuzzy numbers $\tilde{\underline{a}}_i\underline{x}$ and $\tilde{b}_i$.

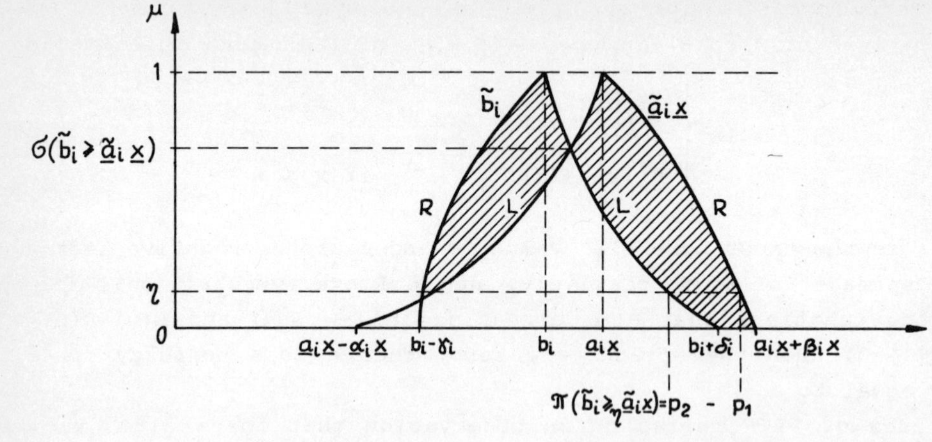

Fig.1 Fuzzy constraint $\tilde{\underline{a}}_i\underline{x} \leq \tilde{b}_i$ for a given $\underline{x}$.

In order to evaluate the degree of possibility for $\tilde{b}_i$ to be greater than or equal to $\tilde{\underline{a}}_i\underline{x}$, 'FLIP' uses two indices, one called optimistic (σ) and another pessimistic (π). The optimistic index $\sigma(\tilde{b}_i \geq \tilde{\underline{a}}_i\underline{x})$ is defined as an ordinate of the intersection point of the right slope of $\tilde{b}_i$ with the left slope of $\tilde{\underline{a}}_i\underline{x}$. Index σ is seen as optimistic because even if it takes value close to 1, the possibility of violation of the constraint $\tilde{\underline{a}}_i\underline{x} \leq \tilde{b}_i$ may be quite big. As a measure of this possibility one can consider the hatched area marked in Fig.1. Thus, in order to make the comparison more credible, one should use σ conjointly with the pessimistic index π.

Pessimistic index π follows from the comparison of the right slopes of $\tilde{\underline{a}}_i\underline{x}$ and $\tilde{b}_i$ at some level $0 \leq \eta \geq 1$. Specifically, $\pi(\tilde{b}_i \geq_\eta \tilde{\underline{a}}_i\underline{x}) = p_2 - p_1$.

Now, one can admit that $\tilde{b}_i \geq \tilde{\underline{a}}_i\underline{x}$ at credibility levels τ and η if and only if

$$\sigma(\tilde{b}_i \geq \tilde{\underline{a}}_i\underline{x}) \geq \tau \quad \text{and} \quad \pi(\tilde{b}_i \geq_\eta \tilde{\underline{a}}_i\underline{x}) \geq \theta \tag{4}$$

where $\tau, \eta \in [0,1]$ and $\theta \in (-\infty, \infty)$. $\theta \geq 0$ means that for any pair (v,y) such that $v \geq \underline{a}_i\underline{x}$, $y \geq b_i$ and $0 \leq \mu_{\underline{a}_i\underline{x}}(v) = \mu_{b_i}(y) \leq \eta$, inequality $y \geq v$ is true. A negative value of θ makes possible that inequality $y \geq v$ is not true.

The constants τ, η and θ are called "safety parameters" because they are responsible for the safety of the assertion that $\tilde{b}_i$ is greater than $\tilde{\underline{a}}_i\underline{x}$. Let us remark that using τ, η and θ one can control the surface of the hatched area marked in Fig.1 which corresponds to the risk of violation of the constraint.

Thus, application of the above comparison principle, transforms every fuzzy constraint (2) into two linear constraints corresponding to conditions (4) on σ and π, respectively (cf.Słowiński, 1990):

$$\underline{a}_i\underline{x} - b_i \le L^{-1}(\tau_i)(\underline{\alpha}_i\underline{x} + \delta_i) \qquad i=1,\dots,m \qquad (5)$$

$$b_i + \delta_i L^{-1}(\eta_i) - \underline{a}_i\underline{x} - \underline{\beta}_i\underline{x}R^{-1}(\eta_i) \ge \theta_i \qquad i=1,\dots,m \qquad (6)$$

In the case of fuzzy objective functions, σ can be used to evaluate the degree of consistency between fuzzy objectives and fuzzy goals. Let the fuzzy cost coefficients and fuzzy goals be:

$$\tilde{\underline{c}}_h = (\underline{c}_h, \underline{\varepsilon}_h, \underline{\kappa}_h)_{LR}, \qquad \tilde{g}_h = (g_h, \phi_h, \nu_h)_{RL}, \qquad h=1,\dots,k$$

Here again, the equality of the reference functions is not a necessary assumption. For any $\underline{x}$, the components of the h-th objective function can be summarized with the extension principle to the fuzzy number:

$$\tilde{\underline{c}}_h\underline{x} = (\underline{c}_h\underline{x},\ \underline{\varepsilon}_h\underline{x},\ \underline{\kappa}_h\underline{x})_{LR}, \qquad h=1,\dots k.$$

Fig.2 shows relative positions of fuzzy numbers $\tilde{\underline{c}}_h\underline{x}$ and $\tilde{g}_h$.

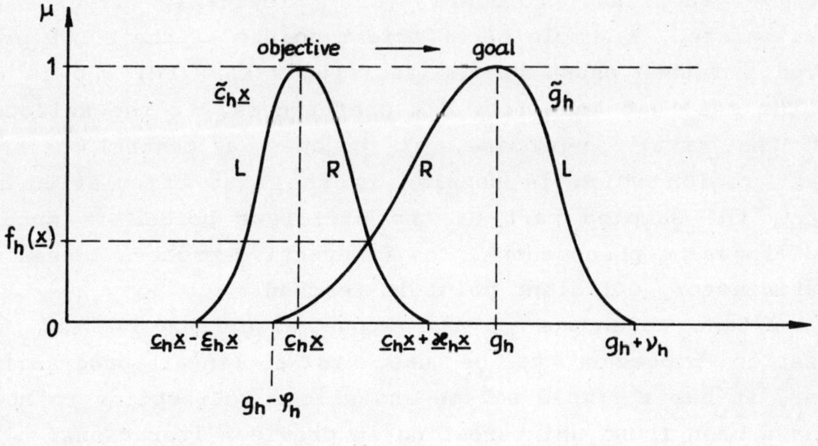

Fig.2 Fuzzy objective $\tilde{\underline{c}}_h\underline{x}$ and fuzzy goal $\tilde{g}_h$ for a given $\underline{x}$.

In order to maximize the consistency between $\tilde{\underline{c}}_h\underline{x}$ and $\tilde{g}_h$ one has to maximize the ordinate $f_h(\underline{x})$ of the intersection point of the right slope of $\tilde{\underline{c}}_h\underline{x}$ with the left slope of $\tilde{g}_h$. Mathematically,

$$f_h(\underline{x})=R[(\underline{c}_h\underline{x} - g_h)/(\kappa_h\underline{x} + \phi_h)], \quad h=1,\dots,k \tag{7}$$

If reference function R of fuzzy cost coefficients and goals is linear (or piecewise linear), then (7) takes the fractional form:

$$f_h(\underline{x})=1 - (\underline{c}_h\underline{x} - g_h)/(\kappa_h\underline{x} + \phi_h), \quad h=1,\dots,k \tag{8}$$

In consequence, for given goals and safety parameters, the fuzzy MOLP problem (1)-(3) is transformed into the following deterministic multiobjective programming problem:

$$\text{Maximize} \begin{bmatrix} f_1(\underline{x}) \\ \dots\dots \\ f_k(\underline{x}) \end{bmatrix} \tag{9}$$

s.t. (3), (5) and (6)

If $f_h(\underline{x})$ are defined by (8), this problem is a multiobjective linear fractional programming (MOLFP) one.

The associate deterministic MOLFP problem is solved using an interactive sampling procedure (cf. Słowiński, 1990). In each calculation step, a sample of efficient points of the MOLFP problem is generated and then shown to the decision maker (DM) who is asked to select the one that best fits his preferences. If the selected point is not the final compromise, it becomes a central point of an efficient region which is sampled in the next calculation step. In this way, the sampled part of the efficient border is successively reduced (focusing phenomenon). The interactive process ceases when the most satisfactory efficient point is reached.

An important advantage of the above algorithm is that the only optimization procedure to be used is a linear programming one. Moreover, it has a simple scheme and allows retractions to the points which have been found uninteresting in previous iterations.

3. INTERACTIVE STEPS OF 'FLIP'

'FLIP' can be summarized in the following steps:

Step 1. Formulation of problem (1)-(3) and definition of fuzzy coefficients.

Step 2. Definition of fuzzy aspiration levels $\tilde{g}_h$ (h=1,...,k) on objectives.

Step 3. Definition of safety parameters τ_i, η_i and θ_i(i=1,...,m) for fuzzy constraints.

Step 4. Formulation of the associate multiobjective deterministic problem (9),(3),(5),(6).

Step 5. Application of an interactive method for solving the associate deterministic problem formulated in step 4.

Step 6. If a best compromise solution has been found then stop, otherwise return to step 3 for revision of safety parameters. Retraction to steps 1 and 2 for redefinition of fuzzy coefficients and/or aspiration levels is also possible.

The DM intervenes in two steps of FLIP. First, when fixing the safety parameters (step 3), and then in the course of the guided generation and evaluation of the efficient points of the associate deterministic problem (step 5). Thus, the interaction with the DM takes place at two levels. As to the first one (step 3), it is worth noting that there are two practical ways of controlling the safety of solutions using parameters τ_i, η_i and θ_i:

(a) fix $\eta_i = 0$, i=1,...,m, and control the optimistic safety with τ_i, and the pessimistic safety with θ_i, i=1,...,m, or

(b) fix $\theta_i = 0$, i=1,...,m, and control the optimistic safety with τ_i, and the pessimistic safety with η_i, i=1,...,m.

The safety parameters are defined taking into account their interval of variation and the knowledge acquired in preceding iterations about the dependency between safety and the quality of the compromise among criteria. If way (a) is chosen, 'FLIP' provides the DM with an information about an approximate interval of variation of Θ_i at level η_i=0 (i=1,...,m).

In the microcomputer implementation of 'FLIP', the efficient points proposed to the DM are presented both numerically, in terms of $\underline{x}$ and middle values of $\tilde{z}_h(\underline{x})$, h=1,...,k, and graphically, in terms of mutual positions of fuzzy numbers corresponding to objectives and aspiration levels on the one hand, and to left- and right-hand-sides of constraints, on the other hand. In this way, the DM gets quite a complete idea about the quality of each proposed solution. The quality is evaluated taking into account the following characteristics:

- scores of fuzzy objectives in relation to the goals,

- dispersion of values of the fuzzy objectives due to uncertainty,
- safety of the solution or, using a complementary term, the risk of violation of the constraints (cf. the hatched areas in Fig.1).

So, the definition of the best compromise involves an analysis of the compromise among criteria and an evaluation of the safety of the corresponding solution.

The analysis of the above characteristics needs indeed a graphical display of objectives and constraints in a form similar to Figs. 1 and 2. We claim that the comparison of fuzzy left- and right-hand-sides of the constraints, as well as evaluation of dispersion of the values of objectives, is practically infeasible on the basis of one or two numerals. The graphical representation of proposed solutions is not only a "user's friendly" interface but the best way of a complete characterization of these solutions.

We put emphasis on this aspect of visual interaction because it is understimated in all procedures proposed for solving multiobjective programming problems under uncertainty and imprecision.

4.SOLVING OF A LAND-USE PLANNING PROBLEM

Let us reconsider a simple decision problem of a land-use planning for a town extension in Israel (cf. Werczberger,1984).

The decision maker has at his disposal 160.5 hectares of land put into extension of a town. The decision consists in such a partitionning of this area between different types of use that future demands for housing and services are satisfied.

One distinguishes seven different uses to which land in extension may be allocated:

-land for low density development, at about 50 d.u./ha (dwelling units per hectare);
-land for medium density development, at about 120 d.u./ha (walk-up apartments);
-land for high density development, at about 160 d.u./ha (elevator apartments);
-land for commercial uses, in ha;
-land for local public services including elementary schools, in ha;
-parks and open space, in ha;
-land for religious services and high schools, in ha.

They correspond to decision variables from x_1 to x_7 defining sizes of areas destined for a given type of use.

Before formulating this problem in terms of a model accepted by 'FLIP', let's remind shortly the original formulation given by Werczberger (1984). Werczberger's model has 9 constraints which are divided into two groups:

-4 fixed constraints, i.e. with crisp coefficients;

-5 stochastic goal constraints, where on the LHS there are crisp coefficients and on the RHS there are goals being random variables with joint normal distributions.

The objective is to find a feasible solution $\underline{x}^*$ which maximizes the total probability of achieving the stochastic goal constraints while strictly satisfying the fixed constraints. That leads finally to a nonlinear programming problem with linear constraints.

In order to present the general capacity of our approach, we propose to consider the fixed constraints as fuzzy constraints where, in general, some coefficient are crisp and others are fuzzy numbers of L-R type. Let's describe these constraints in succession.

The sum of areas used in different ways couldn't exceed the area of vacant land, so the first constraint is:

$$\sum_{i=1}^{7} x_i \leq 160.5 \tag{10}$$

Moreover, three qualitative constraints are defined that assure some standards concerning commercial services, local public services, neighbourhood parks and open spaces. The areas, required for uses mentioned above, are defined as functions of the number of families to be served. Because we may use the fuzzy numbers to define particular coefficients, we can handle the real imprecision or "softness" of some coefficients, so natural for the qualitative constraints.

Formally, one can represent the qualitative constraints in the form of the following fuzzy constraints:

-supply of commercial services

$$-\tilde{a}_{21}x_1 - \tilde{a}_{22}x_2 - \tilde{a}_{23}x_3 + \tilde{a}_{24}x_4 \geq \tilde{b}_2 \tag{11}$$

where $\tilde{a}_{21}=(0.15,0.02,0.02)$, $\tilde{a}_{22}=(0.36,0.03,0.03)$,
$\tilde{a}_{23}=(0.48,0.03,0.03)$, $\tilde{a}_{24}=(100,0,0)$, $\tilde{b}_2=(0,5,5)$;

-supply of local public services (elementary schools)

$$-\tilde{a}_{31}x_1 - \tilde{a}_{32}x_2 - \tilde{a}_{33}x_3 + \tilde{a}_{35}x_5 \geq \tilde{b}_3 \quad (12)$$

where $\tilde{a}_{31}=(0.06,0.01,0.01)$, $\tilde{a}_{32}=(0.144,0.016,0.016)$,
$\tilde{a}_{33}=(0.192,0.02,0.02)$, $\tilde{a}_{35}=(1,0,0)$, $\tilde{b}_3=(0,0.05,0.05)$;

-neighbourhood parks and open spaces

$$-\tilde{a}_{41}x_1 - \tilde{a}_{42}x_2 - \tilde{a}_{43}x_3 + \tilde{a}_{46}x_6 \geq \tilde{b}_4 \quad (13)$$

where $\tilde{a}_{41}=(0.075,0.03,0.03)$, $\tilde{a}_{42}=(0.18,0.04,0.04)$,
$\tilde{a}_{43}=(0.24,0.05,0.05)$, $\tilde{a}_{46}=(1,0,0)$, $\tilde{b}_4=(0,0.05,0.05)$.

We propose, moreover, to transform the LHS's of the stochastic goal constraints into objective functions (criteria), which allow the DM to evaluate the quality of solutions in relation to the goals. The coefficients of the objective functions may also be either crisp or L-R type fuzzy numbers. So, one gets 5 objective functions with (possibly) fuzzy cost coefficients:

-total demand for housing

$$z_1 = \tilde{c}_{11}x_1 + \tilde{c}_{12}x_2 + \tilde{c}_{13}x_3 \quad (14)$$

-housing demand by young couples

$$z_2 = \tilde{c}_{22}x_2 + \tilde{c}_{23}x_3 \quad (15)$$

-housing demand by middle-class families

$$z_3 = \tilde{c}_{31}x_1 + \tilde{c}_{33}x_3 \quad (16)$$

-housing demand by religious families

$$z_4 = \tilde{c}_{41}x_1 + \tilde{c}_{42}x_2 \quad (17)$$

-demand for religious services and facilities

$$z_5 = x_7 \quad (18)$$

where $\tilde{c}_{11}=\tilde{c}_{31}=\tilde{c}_{41}=(50,5,5)$, $\tilde{c}_{12}=\tilde{c}_{22}=\tilde{c}_{42}=(120,12,12)$,
$\tilde{c}_{13}=\tilde{c}_{23}=\tilde{c}_{33}=(160,16,16)$.

Let's precise that all reference functions of fuzzy coefficients appearing in constraints (11)-(13) and objectives (14)-(17) are linear functions.

The next stage of building a model accepted by 'FLIP' is the definition of aspiration levels for different objectives. In Werczberger's model the goals are on the RHS of the stochastic constraints and they are expressed in terms of a normal distribution. The distributions result from population forecast and some socioeconomic analysis of preferences for housing types, schools and so forth.

According to our approach, the goals are considered as appropriate aspiration levels for particular objectives. The aspiration levels are formulated in terms of L-R type fuzzy numbers with the middle values, left and right spreads, as follows:

for z_1 - $\tilde{g}_1=(12000,3000,3000)$,

for z_2 - $\tilde{g}_2=(3000,3000,3000)$,

for z_3 - $\tilde{g}_3=(8000,3000,3000)$,

for z_4 - $\tilde{g}_4=(4000,3000,3000)$,

for z_5 - $\tilde{g}_5=(6.8,3,3)$,

The shape of membership functions of the goals is shown in Fig.5. They approximate the normal distributions given by Werczberger (1984) using three linear pieces for each reference function.

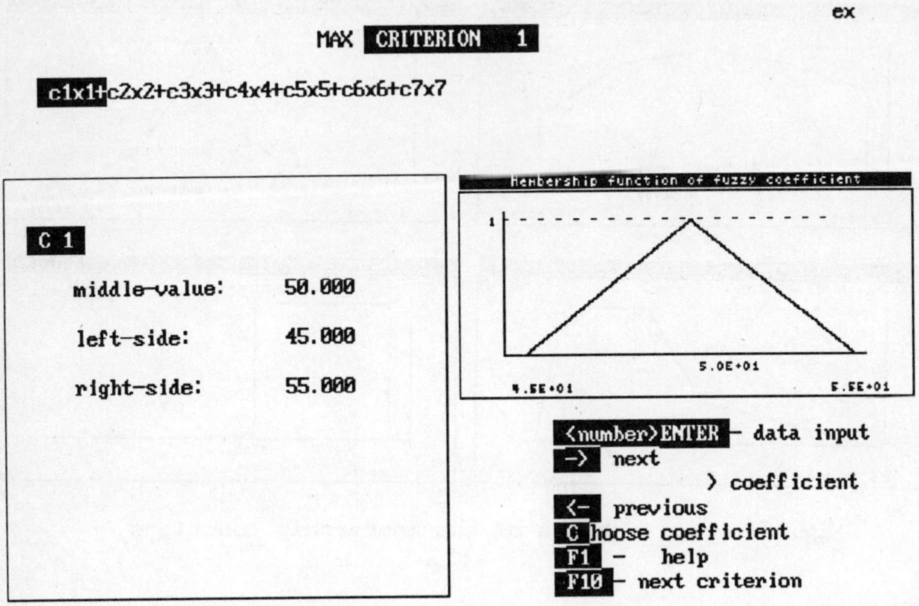

Fig.3 Data input.

The data input to the 'FLIP' package is supported by some graphical facilities. On the monitor screen, the analyst can see at the same time (cf. Fig.3):

- general form of the considered criterion or constraint with a highlighted coefficient which can be defined or changed at this time;
- two windows containning an information about the highlighted coefficient; in the left window, numerical, and in the right window graphical representation of the coefficient is displayed.

'FLIP' allows defining three types of coefficients:

- real numbers - if one wants to have nonfuzzy coefficients (e.g.coefficients of the first constraint in our example);
- linear fuzzy numbers - if one wants to define simple triangular fuzzy numbers (e.g. coefficients of three other constraints);
- piecewise linear fuzzy numbers - if one wants to approximate nonlinear reference functions with up to three linear pieces (e.g. aspiration levels in our example).

Examples of possible shapes of membership functions of a fuzzy number are presented in Fig.4.

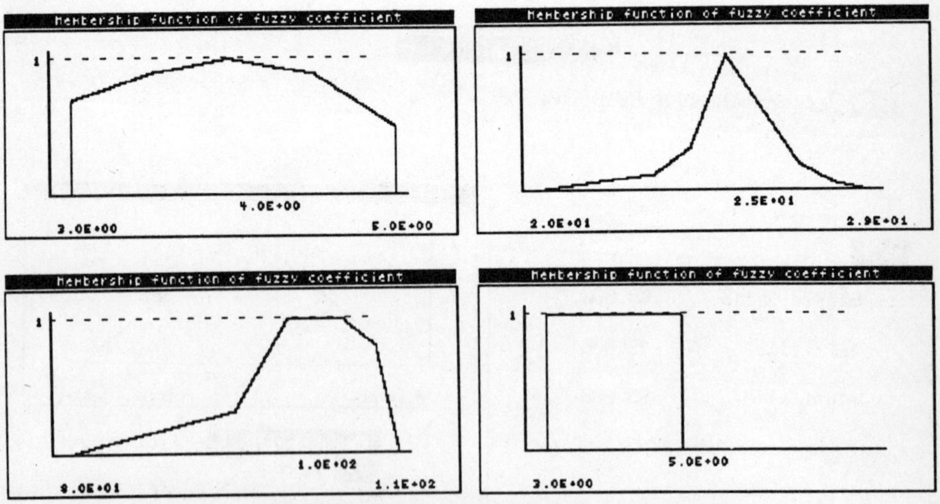

Fig.4 Possible shapes of the membership functions.

The first step of 'FLIP' after the problem formulation consists in defining safety coefficients τ_i, η_i and Θ_i for fuzzy constraints. Let $\tau_i=0.6$, $\eta_i=0$ and $\Theta_i=0$ for all i. Now, the starting point for the sampling method is searched for and immediately after, the interactive part of the procedure starts.

This part consists of successive iterations that are composed of two main steps:

- generation of a sample of efficient solutions to the associate deterministic problem;
- evaluation of the sample by the DM.

In the second step, the generated solutions are presented to the DM both, graphically and numerically. At this moment the DM can see simultaneously four windows, in which the graphical representation of criteria or constraints and values of decision variables may be displayed. He can compare the state of criteria or constraints and see values of variables for four efficient solutions simultaneously.

He has at his disposal the menu that allows him, among others :

- setting display in the highlighted window; it can be the state of a selected criterion or constraint, or values of decision variables for a given solution;
- enlarging the highlighted window to have a better insight into graphical representation of fuzzy criteria or constraints;
- setting numerical display of middle values of criteria; it gives to the DM the first impression about the generated sample.

Before passing to the interactive part of the searching process let's take the denotation: S - cardinality of the generated sample in a given iteration; l_j - l-th solution in the j-th iteration.

Table 1

SOLUTIONS	3_1	4_1	6_1	7_1	8_1	9_1	10_1
z_1	9697	11007	14337	14337	14337	13062	7382
z_2	9697	11007	14337	14337	14337	13062	7382
z_3	1990	9144	10922	10922	10922	3967	3967
z_4	7706	1863	3414	3414	3414	9095	3414
z_5	53	53	18	18	18	18	83

SOLUTIONS	11_1	12_1	13_1	14_1	15_1
z_1	17055	17055	17055	13930	0
z_2	17055	17055	17055	13930	0
z_3	17055	17055	17055	0	0
z_4	0	0	0	13930	0
z_5	0	0	0	0	160.5

Iteration 1

Taking S=15, the DM gets 12 solutions for which middle values of criteria are presented in Tab.1. A tentative of getting 3 other solutions was unsuccessful because the corresponding problems were contradictory. The DM estimates that the most interesting solutions obtained at this stage are 6_1, 7_1, 8_1 (the same point in the objective space). The values of criteria z_1, z_2, z_3 and z_5 at this point are much better then their aspiration levels.

Solution 6_1 is chosen as a new starting point for the search of solutions with a higher value of criterion z_4.

Iteration 2

For S=15 the DM gets 15 solutions (Tab.2). The best solution obtained in this iteration is solution 9_2.

Table 2

SOLUTIONS	1_2	2_2	3_2	4_2	5_2	6_2	7_2	8_2
z_1	14337	14337	14337	14337	14337	15978	15978	15978
z_2	14337	14337	14337	14337	14337	15978	15978	15978
z_3	10922	10922	10922	10922	10922	14531	14531	14531
z_4	3414	3414	3414	3414	3414	1446	1446	1446
z_5	18	18	18	18	18	7.1	7.1	7.1

SOLUTIONS	9_2	10_2	11_2	12_2	13_2	14_2	15_2
z_1	15070	13677	17055	17055	17055	15453	13037
z_2	15070	13677	17055	17055	17055	15453	13037
z_3	9574	12230	17055	17055	17055	8311	13037
z_4	5496	1446	0	0	0	7141	0
z_5	7.1	28	0	0	0	0	38

As we can see in Fig.5, middle values of fuzzy criteria for solution 9_2 are better then middle values of fuzzy aspiration levels. The non-fuzzy criterion z_5 has a better value (7.08) then the middle value of aspiration level (6.8).

So, the DM can say that, from point of view of the values of criteria, this is a satisfactory solution.

To get a full evaluation of solution 9_2, he wants to analyze the state of fuzzy constraints (Fig.6).

In the upper window of Fig.6, we have the constraint corresponding to supply of commercial services. We have a relatively big risk of violation of this constraint, because the great part of the RHS (b) is on the left side of the LHS (a). For other fuzzy constraints (lower windows) the risk of their violation is rather small.

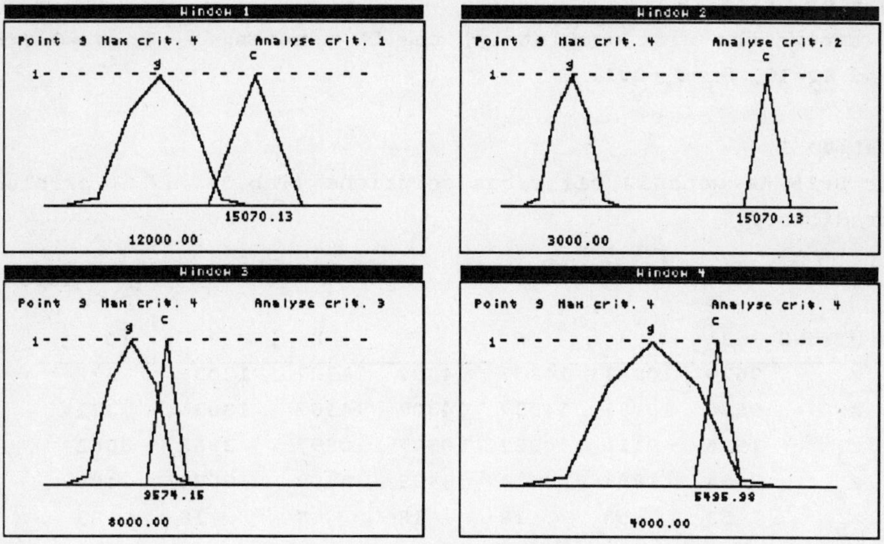

Fig.5 Values of criteria z_1, z_2, z_3 and z_4 for solution 9_2.

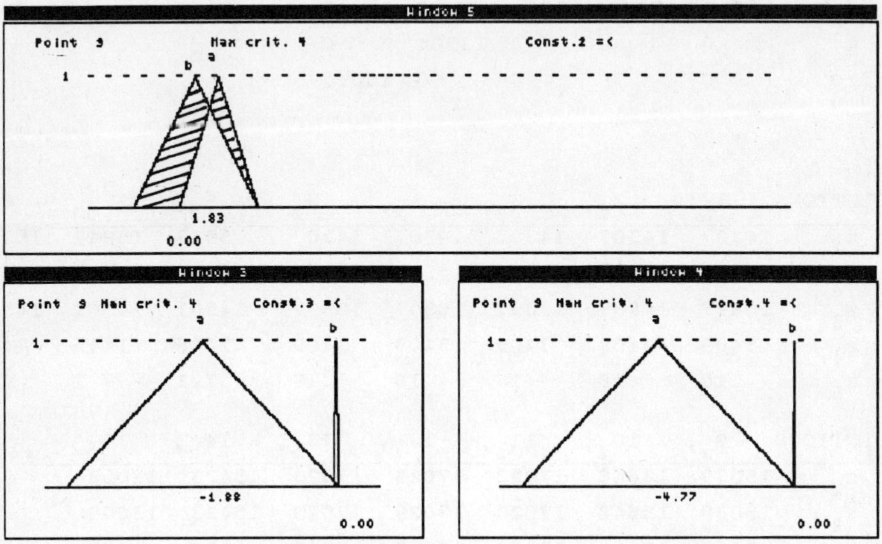

Fig.6 Fuzzy constraints for solution 9_2.

As the values of criteria are very good and there is a big risk of violation of one constraint, the DM wants to remove this risk. It means that the requirements on the safety of constraints should be reinforced but, in consequence, it may lead to some losses on the values of criteria.

To remove the risk completely, the DM assumes: $\tau_i=1$, $\eta_i=0$ (for all i) and $\Theta_2=10$, $\Theta_3=\Theta_4=0.1$.

Iteration 1'

For S=15 he gets 12 efficient solutions (Tab.3) and 3 problems are contradictory.

Table 3

SOLUTIONS	$3_{1'}$	$4_{1'}$	$6_{1'}$	$7_{1'}$	$8_{1'}$	$9_{1'}$	$10_{1'}$
z_1	9670	10981	14337	14307	14307	13036	7372
z_2	9670	10981	14337	14307	14307	13036	7372
z_3	1976	9130	10922	10897	10897	3962	3962
z_4	7694	1851	3414	3409	3409	9074	3409
z_5	53	53	18	18	18	18	83

SOLUTIONS	$11_{1'}$	$12_{1'}$	$13_{1'}$	$14_{1'}$	$15_{1'}$
z_1	17023	17023	17023	13904	0
z_2	17023	17023	17023	13904	0
z_3	17023	17023	17023	0	0
z_4	0	0	0	13904	0
z_5	0	0	0	0	160.5

Table 4

SOLUTIONS	$1_{2'}$	$2_{2'}$	$3_{2'}$	$4_{2'}$	$5_{2'}$	$6_{2'}$	$7_{2'}$	$8_{2'}$
z_1	14307	14307	14307	14307	14307	15947	15947	15978
z_2	14307	14307	14307	14307	14307	15947	15947	15978
z_3	10897	10897	10897	10897	10897	14501	14501	14531
z_4	3409	3409	3409	3409	3409	1445	1445	1445
z_5	18	18	18	18	18	7.1	7.1	7.1

SOLUTIONS	$9_{2'}$	$10_{2'}$	$11_{2'}$	$12_{2'}$	$13_{2'}$	$14_{2'}$	$15_{2'}$
z_1	15039	13648	17023	17023	17023	15423	13008
z_2	15039	13648	17023	17023	17023	15423	13008
z_3	9550	12202	17023	17023	17023	8289	13008
z_4	5489	1445	0	0	0	7133	0
z_5	7.1	28	0	0	0	0	38

Iteration 2'

Taking solution 6_1 as a new starting point, for S=15 the DM gets 15 compromise solutions presented in Tab.4.

The most interesting is solution $9_{2'}$, with a slightly worse values of criteria then solution 9_2.

To complete the evaluation of this solution the DM estimates the state of fuzzy constraints. As we can see in Fig.7, there is no risk of violation of fuzzy constraints.

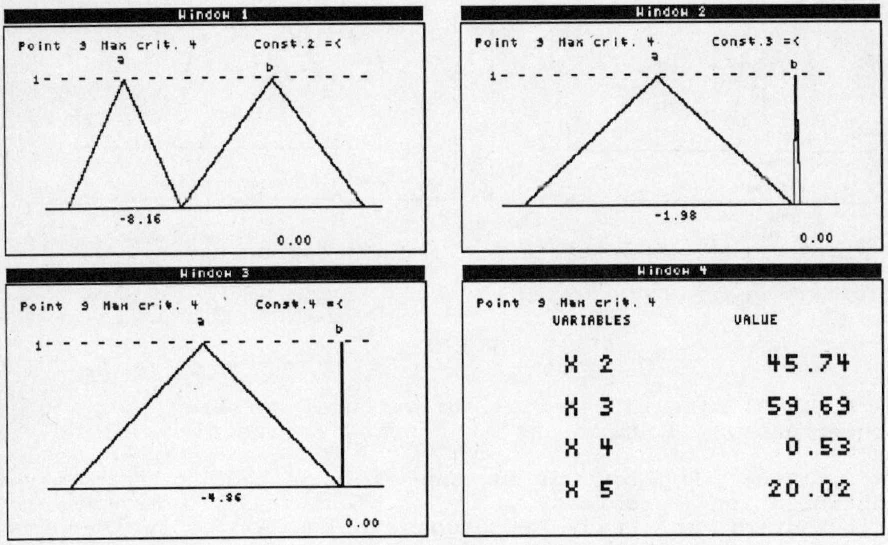

Fig.7 Fuzzy constraints for solution $9_{2'}$.

Finally, the DM accepts solution $9_{2'}$ as the best compromise. The values of decision variables for this solution are the following:

- medium density development	45.74 ha,
- high density development	59.69 ha,
- commercial land	0.53 ha,
- local public services	20.02 ha,
- parks and open spaces	27.42 ha,
- religious services	7.08 ha

and states of fuzzy criteria are presented in Fig.8.

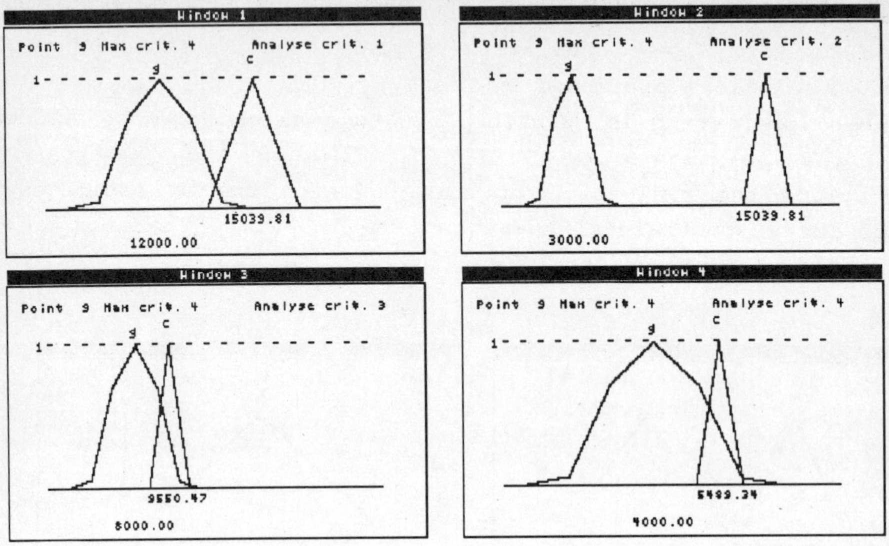

Fig.8 Values of criteria 1,2,3 and 4 for solution 9_2'.

R E F E R E N C E S

Czyżak,P.(1989) 'Multicriteria agricultural problem solving under uncertainty', Foundations of Control Engineering Vol.14, No.2., 61-80.

Czyżak,P.(1990) 'Application of the FLIP method to farm structure optimization problems', in R.Slowinski, J.Teghem (eds.), Stochastic vs. Fuzzy Approaches to Multiobjective Mathematical Programming under Uncertainty, Kluwer Academic Publishers, Dodrecht, 263-278.

Czyżak,P. Słowiński,R.(1989) 'Multiobjective diet optimization problem under fuzziness', in J.L.Verdegay, M.Delgado (eds.), The Interface between Artificial Intelligence and Operations Research in Fuzzy Environment, Verlag TUV Rheinland, Köln, 85-103.

Czyżak,P. Słowiński,R.(1990) 'Solving Multiobjective Diet Optimization Problem under Uncertainty', Proc. of the IIASA International Conference on Multiple Criteria Decision Making, Helsinki, August 1989, Springer Verlag, Berlin (in press).

Słowiński, R. (1986) 'A multicriteria fuzzy linear programming method for water supply system development planning', Fuzzy Sets and Systems 19, 217-237.

Słowiński, R. (1987) 'An interactive method for multiobjective linear programming with fuzzy parameters and its application to water supply planning',in J.Kacprzyk, S.A.Orlovski (eds.), Optimization Models using Fuzzy Sets and Possibility Theory, D.Reidel, Dordrecht, pp.396-414.

Słowiński, R. (1990) '"FLIP": An interactive method for multiobjective linear programming with fuzzy coefficients', in R.Slowinski, J.Teghem (eds.), Stochastic vs. Fuzzy Approaches to Multiobjective Mathematical Programming under Uncertainty, Kluwer Academic Publishers, Dordrecht, 249-262.

Słowiński, R., Teghem, J. (1988) 'Fuzzy Versus Stochastic Approaches to Multicriteria Linear Programming under Uncertainty', Naval Research Logistics, Vol.35, 673-695.

Słowiński, R., Teghem, J.,eds. (1990), Stochastic vs. Fuzzy Approaches to Multiobjective Mathematical Programming under Uncertainty, Theory and Decision Library, ser.D, Kluwer Academic Publishers, Dordrecht.

Słowiński, R., Urbaniak, A., Węglarz, J. (1987) 'Probabilistic and fuzzy approaches to capacity expansion planning of a water supply system', in L.Valadares Tavares, J.Evaristo da Silva (eds.), Systems Analysis Applied to Water and Related Land Resources, Pergamon Press, Oxford, pp.93-98.

Werczberger,E.(1984) 'Planning in an uncertain environment: stochastic goal programming using the versatility criterion', Socio.-Econ.Plan.Sci. Vol.18, No.6, pp.391-398.

FPLP - A PACKAGE FOR FUZZY AND PARAMETRIC LINEAR PROGRAMMING PROBLEMS

Stefan Chanas, Dorota Kuchta and Zenon Nowak

Institute of Industrial Engineering and Management
Technical University of Wrocław
Wybrzeże Wyspiańskiego 27
50-270 Wrocław, Poland

Abstract: An integrated software package designed for solving and analyzing fuzzy and parametric linear programming problems is presented.

Keywords: fuzzy programming, parametric programming, transportation problem, maximum flow problem, minimum cost flow problem.

1. INTRODUCTION

The FPLP (Fuzzy and Parametric Linear Problems) is a system designed for solving and analyzing problems of fuzzy and parametric linear programming. The following problems have been included in the system: the general linear programming problem, the transportation problem, the maximum flow problem and the minimum cost flow problem. Each of these problems is present in the system in two versions: the parametric version and the fuzzy one, i.e. as a problem of parametric programming and as a problem of fuzzy programming.

The system is menu driven and has a context - sensitive help. Several utility functions are provided which permit (in a user - friendly manner) to:

- enter or edit the data for real problems (a full data screen editor),
- save and reload problems from a disk(ette),
- print the data and the solutions.

The system runs on the IBM-PC/XT/AT or equivalent, under DOS 2.1 or higher and needs 512KB of memory. The program is distributed as an executable module on two 360KB double - sided diskettes.

2. Optimization problems included in the FPLP system

2.1 Parametric problems

The FPLP system makes it possible to solve the following problems of parametric linear programming:

(i) The general linear programming problem:

$$\sum_{j=1}^{n} c_j x_j \rightarrow \max\ (\min) \tag{1}$$

s.t.

$$\sum_{j=1}^{n} a_{ij} x_j \leq (\geq, =)\ b_i^1 + b_i^2 \cdot t, \quad i = 1, 2, \ldots, m \tag{2}$$

$$x_j \geq 0, \quad j = 1, 2, \ldots, n \tag{3}$$

$$\text{for } t \in [t_d, t_u] \tag{4}$$

For solving this problem a parametric version of the classical simplex algorithm has been implemented, according to the description given in Nykowski (1980).

(ii) The transportation problem with the parameter in the demand and supply values:

$$\sum_{i=1}^{m} \sum_{j=1}^{n} c_{ij} x_{ij} \rightarrow \min \tag{5}$$

s.t.

$$\sum_{j=1}^{n} x_{ij} = a_i^1 + a_i^2 \cdot t, \quad i = 1, 2, \ldots, m \tag{6}$$

$$\sum_{i=1}^{m} x_{ij} = b_i^1 + b_i^2 \cdot t, \quad j = 1, 2, \ldots, n \tag{7}$$

$$x_{ij} \geq 0, \quad i = 1, 2, \ldots, m,\ j = 1, 2, \ldots, n \tag{8}$$

$$\text{for } t \in [t_d, t_u] \tag{9}$$

For solving this problem in the system, the algorithm presented in Dombrowski (1968) is used.

(iii) The maximal flow problem with the parameter in the arc capacities:

$$v \to \max \tag{10}$$

s.t.

$$\sum_{(k,\, j) \in A} x_{ki} - \sum_{(i,\, j) \in A} x_{ij} = \begin{cases} -v & \text{for } i = s, \\ 0 & \text{for } i \in V \setminus \{s, t\}, \\ v & \text{for } i = t, \end{cases} \tag{11}$$

$$0 \le x_{ij} \le a^1_{ij} + a^2_{ij} \cdot t, \quad (i, j) \in A \tag{12}$$

$$\text{for } t \in [t_d, t_u] \tag{13}$$

where V and A denote, respectively, the set of vertices and the set of arcs of a certain network with a distinguished source $s \in V$ and sink $t \in V$.

For solving the problem (iii) the algorithm from Ruhe (1985) is used, with a correction by Kuchta (1990).

(iv) The minimal cost flow problem with the flow value and the lower and upper flow bounds of the arcs dependant on a parameter.

$$\sum_{(i,\, j) \in A} c_{ij} x_{ij} \to \min \tag{14}$$

s.t.

$$\sum_{(k,\, j) \in A} x_{ki} - \sum_{(i,\, j) \in A} x_{ij} = \begin{cases} -(v_1 + v_2 t) & \text{for } i = s \\ 0 & \text{for } i \in V \setminus \{t\} \\ v_1 + v_2 t & \text{for } i = t \end{cases} \tag{15}$$

$$0 \le x_{ij} \le a^1_{ij} + a^2_{ij} \cdot t, \quad (i, j) \in A \tag{16}$$

$$\text{for } t \in [t_d, t_u] \tag{17}$$

This problem is solved in the package with the help of the algorithm presented in Chanas and Machaj (1987), which is a parametric version of the out-of-kilter algorithm.

For each of the mentioned problems the system provides information about successive critical intervals of the parameter t ($[t_0, t_1]$, $[t_1, t_2]$, ..., $[t_{k-1}, t_k]$, where $t_0 \ge t_d$ and $t_k \le t_u$) and about optimal solutions corresponding to these intervals, which are linear functions of the parameter t. It is also possible to get the graph of the optimal value of the goal function as a function of the parameter t (for the problem (ii) of the minimal total transport cost, for the problem (iii) of the maximal flow value, for the problem (iv) of the minimal total flow cost).

2.2 Fuzzy problems [1)]

The FPLP system offers the possibility of solving four problems of fuzzy linear programming. The first one is the problem of fuzzy linear programming in the general form.

(i') The general problem of fuzzy linear programming:

$$f(x) = \sum_{j=1}^{n} c_j x_j \; \tilde{\geq} \; (\tilde{\leq}) \; g \tag{18}$$

s.t.

$$\sum_{j=1}^{n} a_{ij} x_j \; \tilde{\leq} \; (\tilde{\geq}, \tilde{=}) \; b_i, \quad i = 1, 2, \ldots, m \tag{19}$$

$$x_j \geq 0, \quad j = 1, 2, \ldots, n \tag{20}$$

The symbols $\tilde{\leq}$, $\tilde{=}$, $\tilde{\geq}$ in the above model, as well as in the following ones, are to be read "approximately $\geq$, $=$, $\leq$ ", respectively.

Before we formulate the models of the subsequent problems, we will state the concept of the solution of the problem (i'). This concept will remain valid also for the following problems, which are in fact particular cases of the general problem of the fuzzy linear programming

In order to make the meaning of the fuzzy inequalities $\tilde{\leq}$, $\tilde{=}$, $\tilde{\geq}$ more precise, the model (i') in the system FPLP is replaced with a model, which can be conventionally stated in the following form:

$$f(x) = \sum_{j=1}^{n} c_j x_j \; \tilde{\in} \; G \tag{21}$$

s.t.

---

1) In the part of the FPLP package concerning fuzzy problems, the formulations of the problems, the concepts of their solutions, as well as the methods of determining these solutions presented in Chanas (1988,1989) have been used.

$$\sum_{j=1}^{n} a_{ij}x_j \tilde{\in} B_i, \quad i = 1, 2, \ldots, m \tag{22}$$

$$x_j \geq 0, \quad j = 1, 2, \ldots, n \tag{23}$$

where B_i and G are proper fuzzy intervals of L - R type[2], which are to be given by the user, and the relation "a $\tilde{\in}$ A " is read "a belongs to A to the possibly highest degree".

1) $B_i = (-\infty, b_i, 0, \beta_i)_{L-L}$ iff the i - th condition in the problem (i′) is the $\tilde{\leq}$ inequality.

2) $B_i = (\underline{b}_i, \overline{b}_i, \alpha_i, \beta_i)_{L-L}$ iff the i - th condition in the problem (i′) is the equality $\tilde{=}$, $\underline{b}_i$ and $\overline{b}_i$ are such numbers that

   $\underline{b}_i \leq b_i \leq \overline{b}_i$.

3) $B_i = (b_i, \infty, \alpha_i, 0)_{L-L}$ iff the i - th condition in the problem (i′) is the $\tilde{\geq}$ inequality.

Similarly, $G = (g, \infty, \alpha, 0)_{S-S}$ or $G = (-\infty, g, 0, \beta)_{S-S}$.

Hence, all the fuzzy intervals B_i, i = 1, 2, ..., m, called fuzzy constraints, have to be of the same L - L type. The fuzzy interval, G, called the fuzzy goal, can be of a different, S - S type.

For any vector $x = (x_1, \ldots, x_n) \in \mathbb{R}^n$, the values of the functions $\mu_B(\Sigma_{j=1}^{n} a_{ij}x_j)$, i = 1, ..., m, and $\mu_G(\Sigma_{j=1}^{n} c_jx_j)$, constitute a measure of, respectively, the feasibility degree of the solution x in relation to the i - th fuzzy constraint and the optimality degree of the solution x (the degree to which the solution x satisfies the fuzzy goal G).

The following definition of the solution of the problem (i′), called the maximizing solution, is assumed according to the Bellman-Zadeh concept of decision making in a fuzzy environment (see Bellman and Zadeh, 1970):

<u>Definition 1</u>: Any vector $\hat{x}$, at which the membership function

$$\mu_D(x) = \mu_G(f(x)) \wedge \mu_C(x) \tag{24}$$

---

2) The membership function $\mu_B(x)$ of a fuzzy interval $B=(\underline{b},\overline{b},\alpha,\beta)_{L-R}$ is equal to: 1 for $x \in [\underline{b},\overline{b}]$, $L((\underline{b}-x)/\alpha)$ for $x\in(-\infty,\underline{b})$ and $R((x-\overline{b})/\beta)$ for $x\in(\overline{b},\infty)$. L and R are the shape functions , α and β are the spread parameters. If $\alpha=0$ ($\beta=0$), then it is assumed that $\mu_B(x)=0$ for $x<\underline{b}$ ($x>\overline{b}$) (see Dubois and Prade,1978). In the FPLP system only fuzzy intervals with L=R are used.

where

$$\mu_C(x) = \bigwedge_{1 \le i \le m} \mu_B\left(\sum_{j=1}^{n} a_{ij}x_j\right) \tag{25}$$

takes on the maximal value is called the maximizing solution of the problem (i')[3].

For a fixed solution x the value of the function $\mu_C(x)$ determines the degree, to which x satisfies simultaneously all the fuzzy constraints, and $\mu_D(x)$ the satisfaction degree of the constraints and the goal at the same time.

The following problems included in the package are special cases of the problem (i').

(ii') Fuzzy transportation problem:

$$f(x) = \sum_{i=1}^{m} \sum_{j=1}^{n} c_{ij}x_{ij} \le g \tag{26}$$

s.t.

$$\sum_{j=1}^{n} x_{ij} \cong a_i, \; i = 1, 2, \ldots, m \tag{27}$$

$$\sum_{i=1}^{m} x_{ij} \cong b_j, \; j = 1, 2, \ldots, n \tag{28}$$

$$x_{ij} \ge 0, \; i = 1, 2, \ldots, m, \; j = 1, 2, \ldots, n \tag{29}$$

In order to make the meaning of the fuzzy constraints more precise, the user is required to give the parameters of the fuzzy goal $G = (-\infty, g, 0, \beta)_{S-S}$ and the fuzzy constraints $A_i = (\underline{a}_i, \bar{a}_i, \alpha_i^A, \beta_i^A)_{L-L}$, $i = 1, 2, \ldots, m$, and $B_j = (\underline{b}_j, \bar{b}_j, \alpha_j^B, \beta_j^B)_{L-L}$, $j = 1, 2, \ldots, n$ (the conditions $\underline{a}_i \le a_i \le \bar{a}_i$ and $\underline{b}_j \le b_j \le \bar{b}_j$ have to be fulfilled). The problem is then replaced with the following one:

$$f(x) = \sum_{i=1}^{m} \sum_{j=1}^{n} c_{ij}x_{ij} \tilde{\in} G \tag{30}$$

$$\sum_{j=1}^{n} x_{ij} \tilde{\in} A_i, \quad i = 1, 2, \ldots, m \tag{31}$$

3) The symbol $\wedge$ stands for the minimum operation.

$$\sum_{i=1}^{m} x_{ij} \tilde{\in} B_j, \quad j = 1, 2, \ldots, n \tag{32}$$

$$x_{ij} \geq 0, \quad i = 1, 2, \ldots, m, \quad j = 1, 2, \ldots, n \tag{33}$$

(iii') Fuzzy maximum flow problem:

$$f(x) = v \geq v_0 \tag{34}$$

s.t.

$$\sum_{(k, j) \in A} x_{ki} - \sum_{(i, j) \in A} x_{ij} = \begin{cases} -v & \text{for } i = s \\ 0 & \text{for } i \in V \setminus \{s, t\} \\ v & \text{for } i = t, \end{cases} \tag{35}$$

$$0 \leq x_{ij} \leq \tilde{b}_{ij}, \quad (i, j) \in A \tag{36}$$

This problem is in the system replaced with the following one:

$$f(x) = v \tilde{\in} G \tag{37}$$

$$\sum_{(k, j) \in A} x_{ki} - \sum_{(i, j) \in A} x_{ij} = \begin{cases} -v & \text{for } i = s \\ 0 & \text{for } i \in V \setminus \{s, t\} \\ v & \text{for } i = t, \end{cases} \tag{38}$$

$$x_{ij} \tilde{\in} B_{ij}, \quad (i, j) \in A \tag{39}$$

where $G = (v_0, \infty, \alpha, 0)_{S-S}$ and $B_{ij} = (0, b_{ij}, 0, \beta_{ij})_{L-L}$, $(i, j) \in A$.

(iv') Fuzzy minimum cost flow problem:

$$f(x) = \sum_{(i, j) \in A} c_{ij} x_{ij} \leq c_o \tag{40}$$

$$\underline{v} \tilde{\leq} v \tilde{\leq} \overline{v} \tag{41}$$

s.t.

$$\sum_{(k, j) \in A} x_{ki} - \sum_{(i, j) \in A} x_{ij} = \begin{cases} -v & \text{for } i = s \\ 0 & \text{for } i \in V \setminus \{s, t\} \\ v & \text{for } i = t, \end{cases} \tag{42}$$

$$a_{ij} \tilde{\leq} x_{ij} \tilde{\leq} b_{ij}, \quad (i, j) \in A \tag{43}$$

The following problem corresponds to the above one:

$$f(x) = \sum_{(i,\ j)\ \in\ A} c_{ij}x_{ij} \tilde{\in} G \tag{44}$$

$$v \tilde{\in} V \tag{45}$$

$$\sum_{(k,\ j)\ \in\ A} x_{ki} - \sum_{(i,\ j)\ \in\ A} x_{ij} = \begin{cases} -v & \text{for } i = s \\ 0 & \text{for } i \in V \setminus \{s,\ t\} \\ v & \text{for } i = t, \end{cases} \tag{46}$$

$$x_{ij} \tilde{\in} B_{ij},\ (i,\ j) \in A \tag{47}$$

where $G = (-\infty,\ c_0,\ 0,\ \beta_G)_{S-S}$, $V = (\underline{v},\ \overline{v},\ \alpha_v,\ \beta_v)_{L-L}$ and $B_{ij} = (a_{ij},\ b_{ij},\ \alpha_{ij},\ \beta_{ij})_{L-L}$.

The type of the fuzzy goal, as well as the type of the fuzzy constraints, i.e. of the shape functions S(y) and L(y), can be selected by the user from among the following possibilities:

The fuzzy goal (fuzzy constraints) type	Shape function S(y),L(y)
1. Linear	$S(y)\ (L(y)) = \max(0,1-y)$
2. Exponential	$S(y)\ (L(y)) = e^{-py},\ p\geq 1$
3. Power	$\ddagger S(y)\ (L(y)) = \max(0,1-y^p),p\geq 1$
4. Rational	$\ddagger S(y)\ (L(y)) = 1/(1-y^p),p\geq 1$

Apart from the maximizing solution, for each problem it is possible to obtain with the help of the FPLP package fuzzy solutions of two types. These two solution concepts will be formulated in a shortened way. A more detailed description is presented in Chanas (1989).

With each of the problems (i')-(iv') it is possible to associate a certain problem of parametric programming in a form like those presented in section 2.1, the solution $\{x(t)\}_{t\in[t_d,t_u]}$ ($t_d = L^{-1}(1)$ and $t_u = L^{-1}(0)$) of which fulfills the following conditions (see(25))[4]:

$$\mu_C(x(t)) \geq L(t) \tag{48}$$

---

4) The following fact is worth noticing: The procedures of parametric linear programming included in the FPLP package are consequently used not only in the part of the package concerning the parametric optimization problems, but also in the part concerning the fuzzy problems.

and if $\{x(t)\}$ is a set of basic optimal solutions being nonzero vectors, then:[5)]

$$\mu_C(x(t)) = L(t), \quad t \in [t_d, t_u] \tag{49}$$

It makes sense to consider the following concepts of fuzzy solutions of the initial fuzzy linear programs:

Definition 2: The set of solutions of the proper parametric linear program $\{x(t)\}_{t\in[t_d,t_u]}$, together with the membership function $\mu_{sol}(x(t)) = L(t) \wedge \mu_G(f(x(t))$, is called a fuzzy solution of the type 1 of the initial fuzzy linear program.

The vector maximizing the membership function μ_{sol} is a maximizing solution according to definition 1.

Definition 3: A fuzzy interval FO, given by a membership function

$$\mu_{FO}(r) = \begin{cases} L(t) & \text{for } r = f(x(t) \in [f(x(t_d)),\ f(x(t_u))] \\ 0 & \text{otherwise} \end{cases} \tag{50}$$

is called a fuzzy optimal value of the objective function under the fuzzy constraints of the problem.

The membership function μ_{FO} provides a decision maker with the possibility distribution over a set of optimal values of f. This distribution is generated by the fuzzy constraints of the problem.

Definition 4: The intersection $FO \cap G$ is called a fuzzy solution of the type 2 of the initial fuzzy problem. The solution set $\{x(t)\}_{t\in[t_d,t_u]}$ is called a realizing set for the fuzzy optimal solution $FO \cap G$.

The value belonging to $FO \cap G$ to the maximal degree is realized by the maximizing solution according to definition 1. For each of the fuzzy problems (i')-(iv') it is possible to get with the help of the FPLP package the set of solutions $\{x(t)\}_{t\in[t_d,t_u]}$, as well as graphs of the membership functions of the fuzzy solutions of the type 1 and the type 2, i.e. of the function $\mu_{sol}(x(t))$ for $t \in [t_d, t_u]$ and of the function $\mu_{FO\cap G}(r)$ for $r \in [f(x(t_d), f(x(t_u))]$.

5) Except for the parametric out-of-kilter method all the methods quoted in section 2.1 lead to basic optimal solutions.

3. AN EXAMPLE ILLUSTRATING THE OPERATION OF THE FPLP PACKAGE

The operation of the package will be illustrated with the help of the screen outputs obtained during the analysis of the following fuzzy linear program:

$$3x_1 + 4x_2 + 4x_3 \tilde{\geq} 1750$$

s.t.

$$\begin{aligned} 6x_1 + 3x_2 + 4x_3 &\tilde{\leq} 1200 \\ 5x_1 + 4x_2 + 5x_3 &\tilde{\leq} 1550 \\ x_1, x_2, x_3 &\geq 0 \end{aligned} \tag{51}$$

The user is requested to type in all the coefficients and constants occurring in the problem, as well as to select the type of the constrains (e.g. $\leq$) and of the goal (e.g. min). The types of shape functions are also to be selected (linear, exponential, etc); see Fig.1. The signs $\leq$, $\geq$, $=$ on the screen stand for $\tilde{\leq}$, $\tilde{\geq}$, $\tilde{=}$. Subsequently, the user should give the remaining parameters of all the corresponding fuzzy intervals, i.e. G, B_1, B_2 (parameters of the last column filled with the word "no" signal the lack of these data).

The next screen output (Fig. 2) presents the situation in which the parameters of G and B_1 have been inserted and the parameters of B_2 are just being input. In the analyzed example the following values of the parameters were assumed: $G = (1750, \infty, 150, 0)$, $B_1 = (-\infty, 1200, 0, 100)$, $B_2 = (-\infty, 1550, 0, 200)$. The fuzzy goal is of the linear type and the fuzzy constraints of the exponential type.

The realizing set of the fuzzy solution, i.e. the solution of the proper parametric problem associated with the considered fuzzy one, is given in Fig.3.

The fuzzy solutions of both the considered types for the problem (51) are shown graphically in the screen outputs given in Fig.4 and Fig.5.

The user can obtain exact values of the vector x(t), of the goal function f(x(t)), of $\mu_C(x(t))$ (Sdegree) and of $\mu_G(x(t))$ (Pdegree) for any value of the parameter t, in particular for t = T(optim), corresponding to the intersection point of the curves in Fig.4 and Fig.5 (the maximizing solution according to definition 1, see Fig.6)

REFERENCES

Bellman R.E. and Zadeh L.A (1970), Decision making in a fuzzy environment, Management Science 17, 141-164.

Chanas S. (1988), Wybrane problemy badań operacyjnych z rozmytymi parametrami, Monografia, Politechnika Wrocławska, Wrocław.

Chanas S. (1989), Parametric techniques in fuzzy linear programming, in:

The Interface between Artificial Intelligence and Operations Research in Fuzzy Environment, J.L. Verdegay and M. Delgado (eds.), Verlag TUV Rheinland, 105-115.

Chanas S., Machaj A. (1987), A parametric version of the out-of-kilter method and its application to the fuzzy network flow problems, Wissenschaftliche Berichte der Technischen Universitaet Leipzig, Heft 17, 8-18.

Dombrowski R. (1968), Zagadnienie transportowe z parametrycznymi ograniczeniami, Przegląd Statystyczny, R.XV,103-117.

Dubois D. and Prade H. (1987), Operations on fuzzy numbers, Int. J. of Systems Science 6, 613-626.

Kuchta D. (1990), A note on the Paper "Characterization of all optimal solutions and parametric maximal flows in networks" of G.Ruhe ([6]), Optimization (to appear).

Nykowski (1980), Programowanie liniowe, PWE, Warszawa, 245-260.

Ruhe G. (1985), Characterization of all optimal solutions and parametric maximal flows in networks,Optimization 16, 51-61.

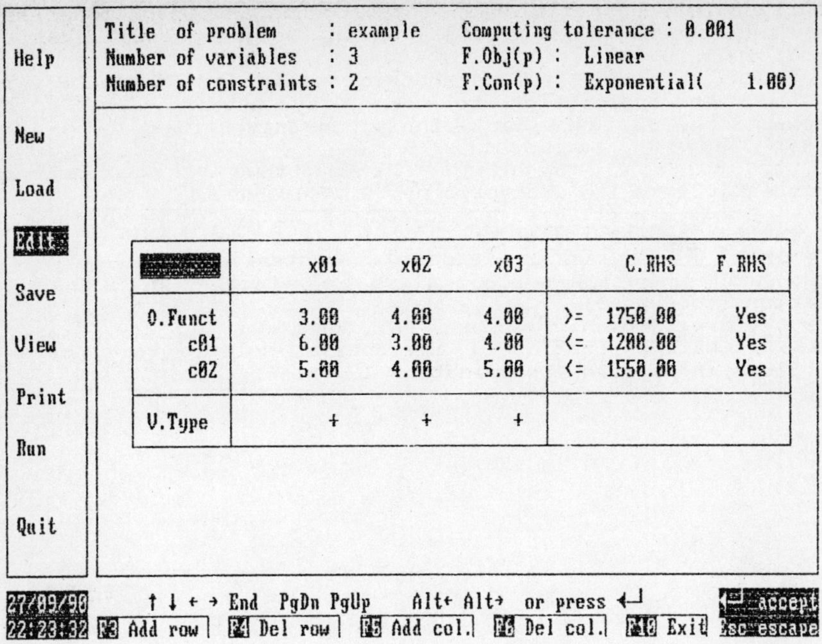

Fig.1: Input of the fuzzy problem (51) without specified parameters of the fuzzy goal and the constraints.

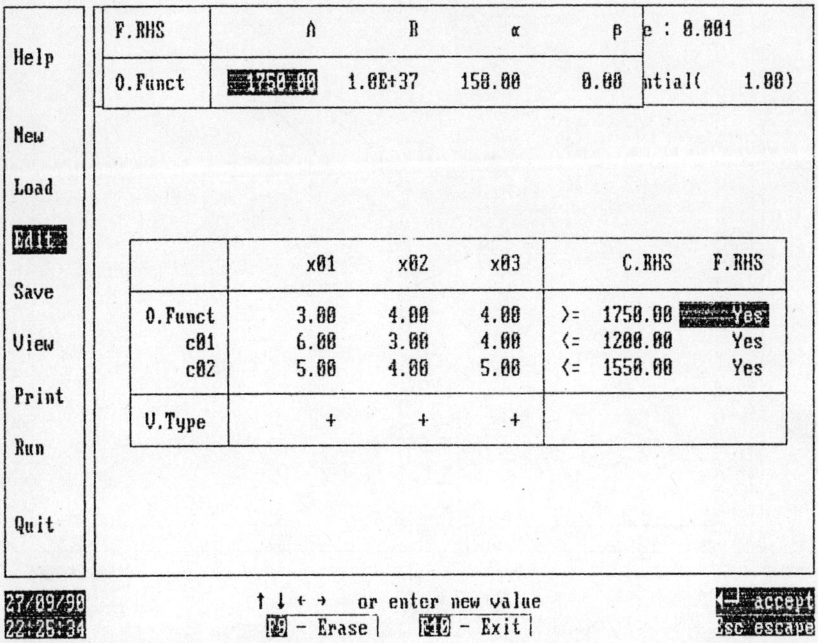

Fig.2: Input of the parameters of the fuzzy constraints and the goal.

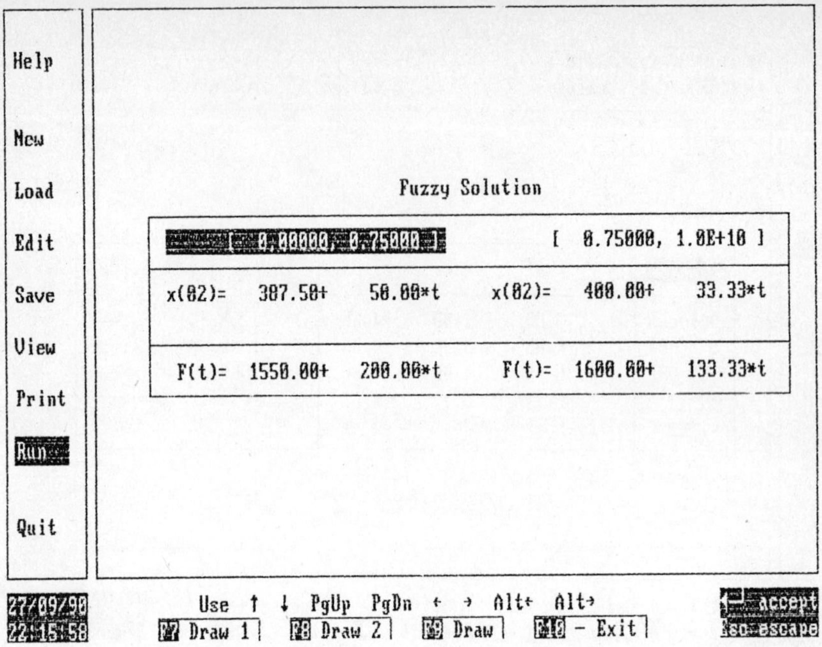

Fig.3: The realizing set of the fuzzy solution for problem (51).

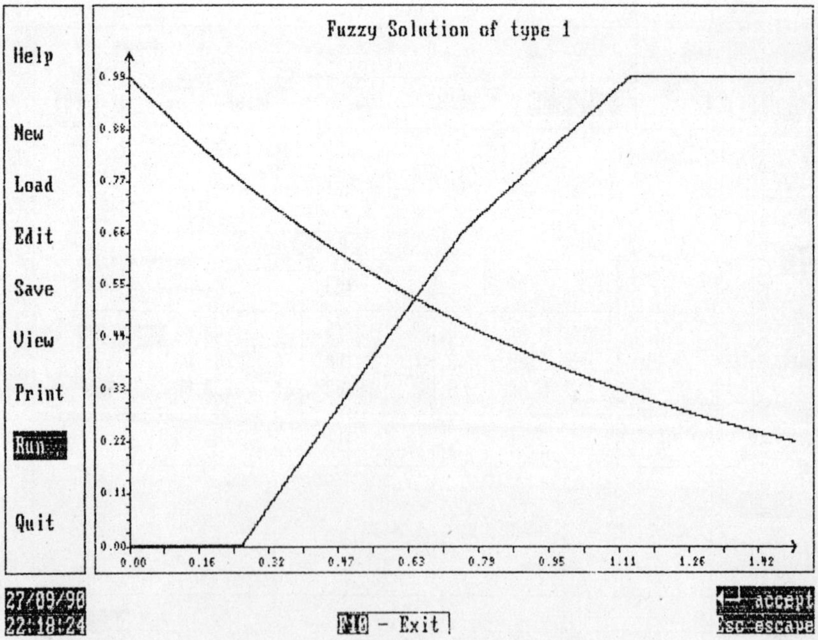

Fig.4: Fuzzy solution of type 1 for problem (51): the ordinate contains the parameter values t, the descending function represents $m_C(x(t))=L(t)$ and the ascending one $m_G(f(x(t))$.

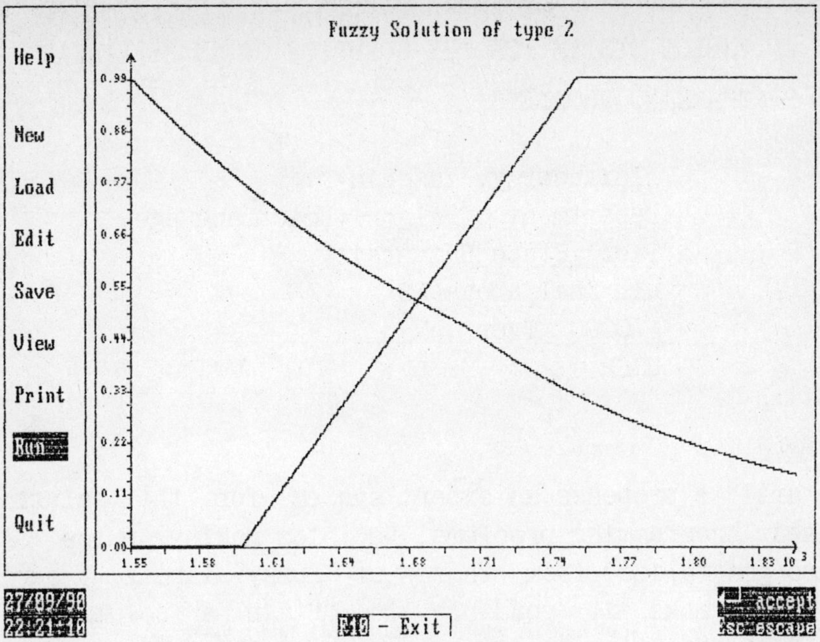

Fig.5: Fuzzy solution of type 2 for problem (51): the ordinate contains the values of the goal function r, the descending function represents the function $m_{F0}(r)$ and the ascending one the function $m_G(r)$;

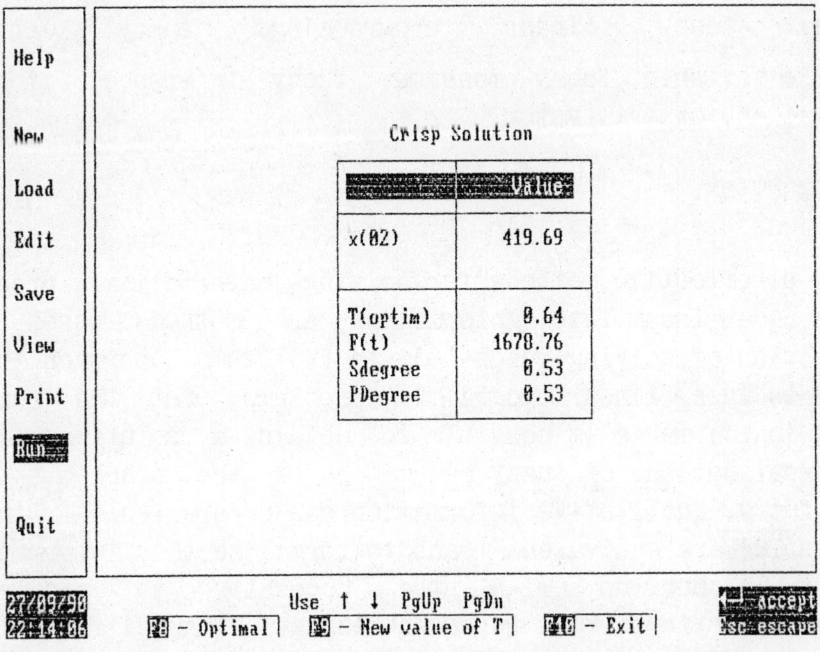

Fig.6: The maximizing solution for problem (51).

AN EXPERT SYSTEM FOR THE SOLUTION OF FUZZY LINEAR PROGRAMMING PROBLEMS

Alexander V. Yazenin
Department of Algorithmic Languages
Tver' State University
ul. Zhelyabova 33
170000 Tver'
USSR

Abstract.

The article proposes an expert system for the solution of fuzzy linear programming problems. Decision making rules of the system are generalized from those of fuzzy linear programming problems and make it possible to obtain a solution under qualitative information conditions. Methods for the estimation of the solving rules matrix are obtained. A fuzzy inference algorithm is developed. An architecture of a expert system is presented. Potential possibilities of the expert system are showed on a model example.
Keywords: fuzzy linear programming, fuzzy variable, linguistic variable, fuzzy measure, fuzzy inference, linguistic approximation, expert system.

1. INTRODUCTION.

One of effective methods for solving mathematical programming problems under incomplete information are methods based on the construction of solving rules [6]. In [4] this approach has been extended to fuzzy linear programming problems. But the rule constructed in [4] makes it possible to obtain a solution only for precise realizations of fuzzy parameters of the problem. In some cases a fuzzy, qualitative information can be available. Linguistic variables [8] are equivalent means to formalize this information. As a result we end up with the necessity to construct a logical-linguistic solving rule. The logical-linguistic solving rule to be discussed is based on the linear rule. The necessity of using experts to estimate the distribution of possible values for indeterminate parameters of the optimization model the implies that the development of its supporting software in the form of a fuzzy expert systems may be expedient.

A model of decision making is considered in section 2.Its formal description is based on the theory of possibility measure [3],[7]. Methods for the estimation of linear solving rule's matrix are proposed in section 3. In section 4 the linear rule is generalized to the case of qualitative information processing. A fuzzy inference (linguistic approximation) algorithm has been developed. An architecture of the expert system developed is given in section 5. Possibilities of the approach are shown by example in section 6.

2. DECISION MAKING MODEL

Let us consider a general procedure for decision making in fuzzy conditions. It is close to the scheme of decision making in stochastic conditions [6]. The essential difference is the following : information about non-deterministic parameters of the fuzzy model is received from experts but is not a result of any objective observation. Stages of information reception and decision making can be interchangeable. If the stage of solving is preceded by the stage of information reception, then a decision making problem has an a priori information structure. In a general case it can be presented by the following chain :

information - decision-making - . . . - decision-making (1)

Unlike stochastic programming the word "observation" has been replaced by the word "information" for the purpose of underlining the principal difference of non-deterministic nature. Information can be precise, fuzzy , received in a qualitative form.

In the conditions of a priori information structure we have to solve a corresponding optimization problem again. This way is not effective. Besides, we often have to obtain a solution immediately after receiving current information. Because of this, it is expedient to construct a solving rule which allows us to make decision in the real time mode.

In correspondence with information structure defined by (1), we will consider a solving rule's constructing method in a linear programming problem :

$$(c,x) \longrightarrow \max, \tag{2}$$

$$\sigma\{ \sum_{j=1}^{n} a_{ij}\, x_j = b_i(\gamma) \} \geqslant \tau_i \ , \ i = 1, \dots , n, \tag{3}$$

$$x \in \mathbf{X}$$

That is a linear programming problem with linear restrictions on the fuzziness. In this problem (c,x) is a scalar product of vectors c, $x \in \mathbf{E}^n$, where $\mathbf{E}^n$ is measurable n-measured Euclidean space, $\mathbf{X} \subset \mathbf{E}^n$ is a set, satisfying restrictions of a special type, $a_{ij} \in \mathbf{E}^1$, $b_i(\gamma)$ are fuzzy (possibilistic) variables , defined in a possibility space $(\Gamma, \mathbf{P}(\Gamma), \sigma)$. In this model of fuzziness [3], $\gamma \in \Gamma$ is a set of elements , $\mathbf{P}(\Gamma)$ is the set of all subsets of Γ , σ is a fuzzy measure, satisfying the following conditions :
1) $\sigma(\emptyset)=0$, $\sigma(\Gamma)=1$; 2) $\sigma(\bigcup_{\alpha} A_\alpha)=\sup \sigma(A_\alpha)$, $\forall\ A_\alpha \in \mathbf{P}(\Gamma)$.

Naturally, as in probability theory, the possibility space $(\Gamma, \mathbf{P}(\Gamma), \sigma)$ is in the background, and distribution functions of possible values for fuzzy parameters of b_i come forward. These functions are determined in the following way :

$$\mu_{b_i}(t) = \sigma\{ \gamma \in \Gamma \mid b_i(\gamma) = t \} , \ \forall\ t \in \mathbf{E}^1$$

i.e. $\mu_{b_i}(t)$ is the possibility that fuzzy variable b_i can take on value t.

If $\mathbf{X}=\mathbf{E}^n_+ = \{ x \in \mathbf{E}^n \mid x \geqslant 0 \}$, then we can give the following interpretation to model (2) and (3). Let the vector $b(\gamma)=(b_1(\gamma),\dots,b_m(\gamma))$ contain possible values of resources (raw products, materials), vector $x=(x_1,\dots,x_n)$ be a vector of manufactured volume, a be an expense volume of resource i for product j , vector $c=(c_1 ,\dots, c_n)$ be a price vector of output units. It is requred to find an output production plan $x=(x_1,\dots,x_n)$ providing the maximum income with the possibility to satisfy a resources expense balance, if the quantity of these resources is not less than threshold values $\tau_i \in (0,1]$, $i = 1,\dots,m$.

Since the model (2) and (3) is linear it is expediently to present a solving rule in the following type :

$$x(\gamma) = D\, b(\gamma) \tag{4}$$

where $D = \{ d_{ij} \}$ is an $n*m$ solving rule matrix with determistic coefficients $d_{ij} \geq 0$.

The functional dependence (4) gives a method for correcting plan x under new information about resources, that does not require to solve (2) and (3). The matrix of the rule (4) can be estimated according to a priori information about the distribution of possible values of vector $b(\gamma)$. Let us proceed to estimation methods of the solving rule matrix .

3. ESTIMATION OF THE SOLVING RULE'S MATRIX

We can take a solution of the following problems as an estimation of the solving rule's matrix (4).

P1. The problem of level maximization :

$$k \longrightarrow \max, \tag{5}$$

$$\begin{cases} \sigma\{ (c, D\, b(\gamma)) = k \} \geq \tau_0 , \\ \sigma\{ (a^i, D\, b(\gamma)) = b_i(\gamma) \} \geq \tau_i , \ i = 1, \dots , m, \\ d_{ij} \geq 0 , \ i = 1, \dots , n, \ j = 1, \dots , m , \end{cases} \tag{6}$$

where $a^i = (a_{i1} , \dots , a_{in})$, $\tau_0 \in (0,1]$ is a given threshold value.

P2. The problem of a modal goal function maximization :

$$\mathbf{M}\{ (c, Db(\gamma)) \} \longrightarrow \max, \tag{7}$$

$$\begin{cases} \sigma\{ (a^i, D\, b(\gamma)) = b_i(\gamma) \} \geq \tau_i , \ i = 1, \dots , m, \\ d_{ij} \geq 0 , \ i = 1, \dots , n, \ j = 1, \dots , m , \end{cases} \tag{8}$$

where **M** denotes the transition to modal values of corresponding fuzzy parameters. This decision making principle is equvalent to the determination of the average for parameters in stochastic programming.

We shall use classes of normal and symmetric triangular fuzzy variables for the modelling of fuzzy parameters:

$$\mu_{b_i}(y) = \exp\{-((y-\alpha_i)/\beta_i)^2\} \in \mathbf{N}(\alpha_i, \beta_i), \ y \in \mathbf{E}^1, \tag{9}$$

$$\mu_{b_i}(y)=\max\{0,\min\{1-c_{b_i}(y-\alpha_i),1+c_{b_i}(y-\alpha_i)\}\}\in\mathbf{Tr}(\alpha_i,\beta_i), y\in\mathbf{E}^1. \tag{10}$$

where α_i is a modal value of the fuzzy variable, β_i is a coefficient of fuzziness, $c_{b_i} = 2/\beta_i$, $i = 1, \dots, m$.

Let us proceed to methods of problems solving (5),(6) and (7),(8). The essence of these methods is the construction of their equivalent precise analogues.

Theorem 1. Let $b_1(\gamma), \dots, b_m(\gamma)$ be unrelated [3] triangular (normal) fuzzy variables with parameters $(\alpha_1,\beta_1),\dots,(\alpha_m,\beta_m)$, respectively. Then the problem (5) and (6) has the following equivalent precise analogue :

$$k \longrightarrow \max, \tag{11}$$

$$\begin{cases} r_0 \leqslant (k - (c, D\alpha)) / (c, D\beta) \leqslant R_0 , \\ r_i \leqslant (\alpha_i - (a^i_; D\alpha)) / (\beta_i + (a^i_; D\beta)) \leqslant R_i, i=1,\dots,m, \\ d_{ij} \geqslant 0 , i = 1, \dots , n, j = 1, \dots , m , \end{cases} \tag{12}$$

where $\alpha = (\alpha_1,\dots,\alpha_m)^T$, $\beta=(\beta_1,\dots,\beta_m)^T$, $r_i= (\tau_i-1)/2$, $R_i= (1-\tau_i)/2$, in the case of triangular fuzzy variables, and r_i and R_i are roots of the equation $\exp(-z^2)-\tau_i = 0$ in the case of normal fuzzy variables, $i = 0, \dots , m$.

The following lemma will be required for providing Theorem 1.

Lemma. The fuzzy variable $q_i=((b_i - \alpha_i)-(a^i_; D(b-\alpha))) / (\beta_i+(a^i_; D\beta$ has the following properties :

a) $\mu_{q_i}(y) \in \mathbf{N}(0,1)$ if the fuzzy variables $b_i(\gamma)$ are unrelated and $\mu_{b_i}(y) \in \mathbf{N}(\alpha_i,\beta_i)$, $i = 1, \dots , m$;

b) $\mu_{q_i}(y) \in \mathbf{Tr}(0,1)$ if the fuzzy variables $b_i(\gamma)$ are unrelated and $\mu_{b_i}(y) \in \mathbf{Tr}(\alpha_i,\beta_i)$, $i = 1, \dots , m$.

Proof. We shall give it for the case of triangular fuzzy variables. Indeed

$$\mu_{q_i}(z)= \mu_{b_i-\alpha_i-(a,D(b-\alpha))}\ (z(\beta_i+(a^i_; D\beta)))=$$

$$\mu_{b_i-(a^i,Db)}\ (z(\beta_i+(a^i_; D\beta)) +\alpha_i-(a^i_; D\alpha))$$

From [3] it follows that the fuzzy variable $b_i-(a^i,Db)$ is a triangular fuzzy variable with parameters $(\alpha_i-(a^i_; D\alpha),\beta_i+(a^i_; D\beta))$. Therefore $\mu_{q_i}(z)=\max\{0,\min\{ 1-c_i z\ (\beta_i+(a^i_; D\beta)),\ 1+c_i(\beta_i+(a^i_; D\beta))\} \}$. Since $c_i = 2/(\beta_i+(a^i_; D\beta))$, then $\mu_{q_i}(z)=\max\{0,\min\{ 1-2z,1+2z \} \}\in \mathbf{Tr}(0,1)$.

Analogously, this lemma can be proved for the class of normal fuzzy variables.

Theorem proof. We will give it according to a scheme proposed in [4]. Let us transform restriction i ($i \geqslant 1$) from system (6) with a fixed matrix D:

$$\sigma \{ (a^i, Db)=b_i \} = \sigma \{ \{b_i-(a^i_{;} Db) -(\alpha_i-(a^i_{;} D\alpha))+ (\alpha_i-(a^i_{;} D\alpha))\}/ (\beta_i+(a^i_{;} D\beta)) = 0 \} = \mu_{q_i} ((\alpha_i-(a^i_{;} D\alpha)) / (\beta_i+(a^i_{;} D\beta)))$$

It follows from the lemma, that $\mu_{q_i}(z) \in \mathbf{Tr}(0,1)$ for any matrix D, i.e. a fuzzy variable q_i is not dependent on D. Therefore, restriction i in the system (6) is equivalent to

$$\mu_{q_i} ((\alpha_i-(a^i_{;} D\alpha)) / (\beta_i+(a^i_{;} D\beta))) \geqslant \tau_i \text{ , or}$$

$$r_i \leqslant (\alpha_i-(a^i_{;} D\alpha))/(\beta_i+(a^i_{;} D\beta)) \leqslant R_i \text{ ,}$$

where $r_i = (\tau_i-1)/2$, $R_i = (1-\tau_i)/2$, $i = 1, \ldots, m$.

Now, let us consider the first restriction of system (6).By means of equivalent transformations we have :

$$\sigma\{(c,Db)=k\} = \sigma\{(c,Db)-k=0\} = \sigma\{(c,Db)-(c,D\alpha)+(c,D\alpha)-k=0\} = \sigma\{((c,Db)-(c,D\alpha)+(c,D\alpha)-k)/(c,Db)=0\}.$$

Denote : $q_0=((c,Db)-(c,D\alpha))/(c,D\beta)$. Then $\sigma\{(c,Db)=k\} = \mu_{q_0}((k-(c,D\alpha))/(c,D\beta))$. The lemma is also true for q_0 .

Therefore, the restriction $\sigma\{(c, Db)=k\} \geqslant \tau_0$ is equivalent to

$r_0 \leqslant (k-(c,D\alpha))/(c,D\beta) \leqslant R_0$, where $r_0 = (\tau_0-1)/2$, $R_0 = (1-\tau_0)/2$

In the analogous way a proof can be given for a class of normal fuzzy variables. Q.E.D.

Theorem 2. Let $b_1(\gamma),\ldots,b_m(\gamma)$ be unrelated triangular (normal) fuzzy variables with parameters $(\alpha_1,\beta_1),\ldots,(\alpha_m,\beta_m)$ Then the problem (7) and (8) is equivalent to the following problem

$$(c,D\alpha) \longrightarrow \max, \tag{13}$$

$$\begin{cases} r_i \leqslant (\alpha_i-(a^i_{;}D\alpha))/(\beta_i+(a^i_{;}D\beta)) \leqslant R_i, \; i=1,\ldots,m, & (14) \\ d_{ij} \geqslant 0, \; i=1,\ldots,n, \; j=1,\ldots, m, \end{cases}$$

where $\alpha = (\alpha_1,\ldots,\alpha_m)^T$, $\beta=(\beta_1,\ldots,\beta_m)^T$, $r_i = (\tau_i-1)/2$, $R_i = (1-\tau_i)/2$, in the case of triangular fuzzy variables, and r_i, R_i are roots of the equation $\exp(-z^2)-\tau_i=0$ in the case of normal fuzzy variables, $i = 1, \ldots, m$.

Proof. The system of restrictions (8) is transformed into the equivalent system (14), and the proof follows from that of Theorem 1. $\mathbf{M}\{(c,Db(\gamma))\}=(c,D\alpha)$ follows from a corresponding theorem [3]. Q.E.D.

It is easy to see, that problems (11) and (12) and (13) and (14) can be transformed into standard problems of linear programming.

The variable k and the elements of matrix D in problem (11) and (12), and the elements of matrix D in problem (13) and (14) are variable parameters. We can take a solution D^* of this problems as an estimation of the solving rule's matrix (4).

4. A LOGICAL-LINGUISTIC SOLVING RULE

A functional relation (4), considered in section 2, allows us to obtain the components of vector x which is the solution of problem (2) and (3), using possible values for the components of a fuzzy vector $b(\gamma)$ and not solving the problem again. In addition, the components of this vectors are precise. It is important to generalize this rule to the case, when input information of a system being optimized is in a qualitative form, in particular, in the form of statements of a quasi-natural language. Such information can be formalized by a linguistic variable [8]. Naturally, in this case a solution that is a vector will also have components, which are values of some linguistic variables.

The solving rule (4), under qualitative information, will be called a logical-linguistic solving rule.

Let us proceed to a formal description of the logical-linguistic solving rule and a fuzzy inference procedure by means of the linguistic variable terminology [8].

Let the triplets $\mathbf{B}_i=(B_i,T(B_i),U_{B_i})$, $i=1,\dots,m$ and $X_j=(X_j,T(X_j),U_{X_j})$ $j=1,\dots,n$ describe linguistic variables, corresponding to the input and output parameters. Here B_i,X_j are names of linguistic variables $T(B_i)$ and $T(X_j)$ are term sets of these variables. $U_{B_i},U_{X_j}\subset \mathbf{E}^1$ are universal sets. $T(B_i)$ and $T(X_j)$ form a collection of the fuzzy numbers mentioned above: $T(B_i)=\{b_{i1},\dots,b_{ip}\}$, $T(X_j)=\{x_{j1},\dots,x_{jq}\}$, whose possible values are defined by their coresponding membership functions.

The fuzzy inference will be meant as a procedure for obtaining the output vector whose components are values of the linguistic variables $x^{out}=(x_1^{out}, \dots, x_n^{out})$ induced by the vector of values of linguistic variables $b^{in}=(b_1^{in},\dots,b_m^{in})$, $b_i^{in}\in T(B_i)$, $i=1,\dots,m$; $x_j^{out}\in T(X_j)$, $j=1,\dots,n$. Membership functions, corresponding to the linguistic values b_{ik}, x_{jl}, will be denoted $\mu_{b_{ik}}$ and $\mu_{x_{jl}}$, respectively. The fuzzy inference in this case is equivalent to the solution of a linguistic approximation problem of a fuzzy linear system, i.e.

$$x(\gamma)=Db(\gamma) \tag{15}$$

where vector $b(\gamma)$, corresponding to vector b^{in}, has fuzzy components.

We ought to choose such components x_j of vector x which approximate the fuzzy output of system (15), from the maximum-similarity-measure-type condition for fuzzy variables x_j and x_{jl}. This measure can be defined in the following way :

$$\sigma\{x_j \text{ is } x_{jl}\}= \max_y \sigma\{x_j=y,\ x_{jl}=y\}= \max_y \mu_{x_j,x_{jl}}(y,y) \tag{16}$$

If the fuzzy variables x_j and x_{jl} are unrelated, then

$$\sigma\{x_j \text{ is } x_{jl}\}= \bigvee_y (\mu_{x_j}(y) \wedge \mu_{x_{jl}}(y)) \tag{17}$$

The following theorem determines a fuzzy inference algorithm.

Theorem 3. Let $b_1, \dots, b_m$ be triangular (normal) fuzzy variables with parameters $(\alpha_1,\beta_1),\dots,(\alpha_m,\beta_m)$. Then

$$\mu_{x_i^{out}}= \bigvee_j \bigvee_y (\mu_{x_i}(y) \wedge \mu_{x_{ij}}(y)) \tag{18}$$

where x_i is a triangular (normal) fuzzy variable with parameters $(\sum_{j=1}^m d_{ij}\,\alpha_j, \sum_{j=1}^m d_{ij}\,\beta_j)$, $i=1,\dots,n$.

Proof. To prove this theorem it is sufficient to show that x_i is a triangular (normal) fuzzy variables. Since the system (15) is linear, then $x_i = \sum_{j=1}^m d_{ij}\, b_j$. The variable x_i is a triangular (normal) fuzzy variable, because the families of triangular and normal fuzzy variables are closed under addition operation [3]. It also follows from [3] that x_i has parameters $(\sum_{j=1}^m d_{ij}\,\alpha_j, \sum_{j=1}^m d_{ij}\,\beta_j)$. Q.E.D.

Note. The implementation of fuzzy inference algorithms requires the computing of similarity measures for fuzzy variables. It is easy in the class of triangular fuzzy variables. Using statement 5.4 from [1] (page 119) we can prove the following proposition.

Proposition. Let Q and P be unrelated triangular fuzzy variables with parameters (α_1,β_1) and (α_2,β_2). Then

$$\sigma\{Q \text{ is } P\}= \max\{1-((2(\alpha_1-\alpha_2)/(\beta_1+\beta_2))\vee(2(\alpha_2-\alpha_1)/(\beta_1+\beta_2))\vee 0),0\} \quad (19)$$

5. ARCHITECTURE OF A FUZZY EXPERT PLANNING SYSTEM

The specifies of a subject and requirements as to the expert system determine the composition of its components.

The given expert system must have the following components:

1. Knowledge base.
2. Component of knowledge elicitation.
3. Interaction component.
4. Database.
5. Linear programming package.

Connections between the components of the expert system have been presented in Fig. 1.

The expert system can work in two modes :

a) knowledge elicitation;

b) problem solving.

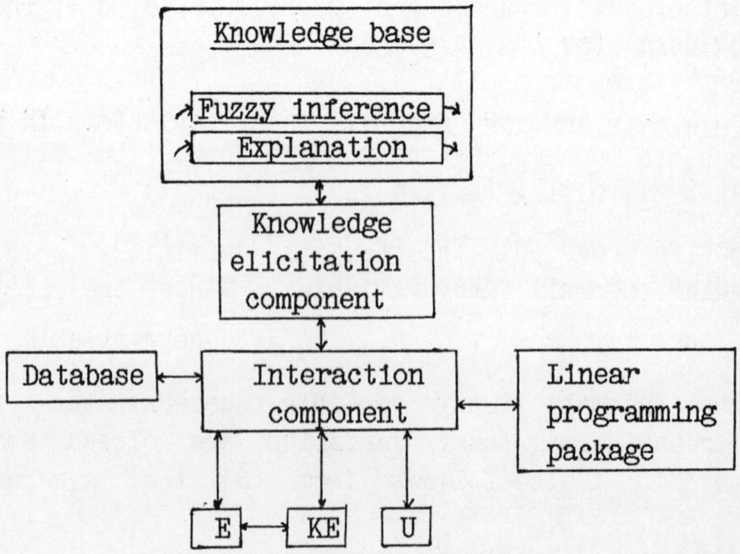

Fig.1.Scheme of the expert system.

In the knowledge elicitation mode an expert (E) with assistance of a knowledge engineer (KE) constructs a knowledge base and a database that is:

- input parameters of the possibility distribution for fuzzy variables $b_1, \dots, b_m$ and threshold values $\tau_0, \dots, \tau_m$;
- forms and inputs the term sets $T(B_i)$, $T(X_j)$ of linguistic variables $B_i, X_j, i=1, \dots, m$; $j=1, \dots, n$;
- performs a synthesis of equivalent precise analogue for the estimation of the solving rule matrix and makes a decision on the base of this analogue .

A knowledge base is built with the help of the interaction and knowledge elicitation components.

It is the user (U) who is interacting with the expert system during the problem solving phase. In the dialogue process the expert system receives current information about the decision making situation (declarative knowledge) from him. In this case this contains the components of the resource vector.

A user query to knowledge base is formulated according to the principle : "Enter current values of the resources and product name - obtain the product output plan and its possibility measure".

6. MODEL EXAMPLE

Let us assume that a company wants to make a daily production plan for three products : x_1, x_2, x_3. The production of the requires to use two resources: b_1, b_2. Possible values of the resources are modelled by fuzzy variables $\mu_{b_1}(y) \in \mathbf{Tr}(4,2)$,$\mu_{b_2}(y) \in \mathbf{Tr}(5,2)$. Let c_j be the unit cost of product j , a_{ij} be the volume of resource i expense for product j, $i=1,2; j=1,2,3$. The plan required may be obtained by the solution of the following problem:

$$c_1x_1+c_2x_2+c_3x_3 \longrightarrow \max,$$

$$\begin{cases} \sigma\{ a_{11}x_1+a_{12}x_2+a_{13}x_3 = b_1(\gamma) \} \geqslant 0.6, \\ \sigma\{ a_{21}x_1+a_{22}x_2+a_{23}x_3 = b_2(\gamma) \} \geqslant 0.5, \\ x_1 . x_2 . x_3 \geqslant 0 \end{cases}$$

Let $c=(2,3,2)$, $a^1=(1,2,2)$, $a^2=(2,1,2)$.
Information about the current values of resources is received in aqualitative form and is modelled by means of linguistic variables:
B_1= Resource_1, $T(B_1)$={big, middle, small},
B_2= Resource_2, $T(B_2)$={big, middle, small},

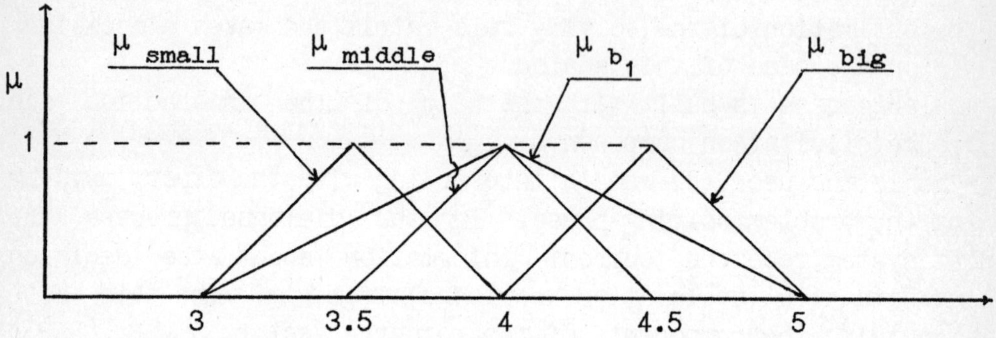

Fig.2. Term set of the linguistic variable "Resource_1".

Fuzzy restrictions on the values of linguistic variable "resource_1" have been specified by triangular variables with parameters (4.5, 1), (4, 1) and (3.5, 1). The term set of this linguistic variable has been shown in Fig.2. Fuzzy restrictions on the values of linguistic variable "resource_2" are also specified by triangular fuzzy variables with parameters (5.5, 1), (5, 1) and (4.5, 1). The term set of this linguistic variable has been represented in Fig.3.

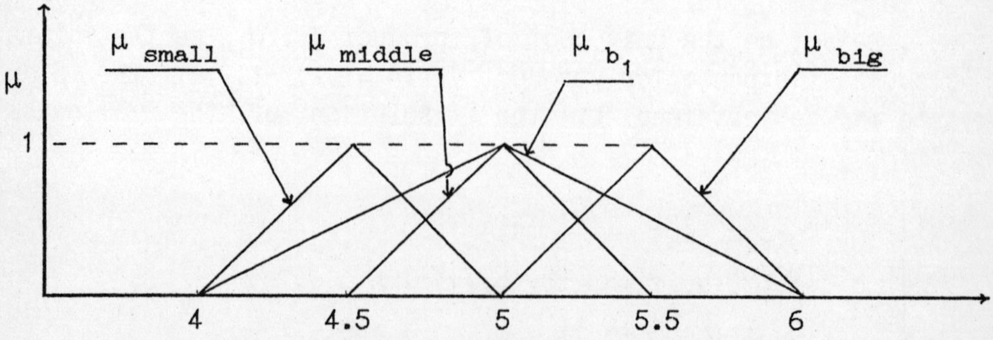

Fig.3. Term set of the linguistic variable "Resource_2".

Possible production plans are modelled by the linguistic variables : X_1="Product_1", X_2 ="Product_2" and X_3="Product_3". These linguistic variables have the same term sets: $T(X_1)=T(X_2)=T(X_3)$ = { much, not_much, little, very_little }. Fuzzy restrictions on the first three values are specified by triangular fuzzy variables with the parameters (3, 2), (2, 2) and (1, 2), while

$$\mu_{very\_little}(y)= \begin{cases} 0, & \text{if } y \notin [0,1], \\ 1-y, & \text{if } y \in [0,1], \end{cases}$$

A term set, corresponding to these linguistic variables, is shown in Fig.4.

For this problem the solving rule matrix has the following form :

$$D = \begin{pmatrix} d_{11} & d_{12} \\ d_{21} & d_{22} \\ d_{31} & d_{32} \end{pmatrix}$$

Let the problem with the threshold value τ_0=0.5 be solved for its estimation. Using Theorem 1 and performing the corresponding computation and transformations we shall get the following linear programming problem :

$$k \longrightarrow \max,$$

$$\begin{cases} 7\,d_{11}+9\,d_{12}+10.5d_{21}+13.5d_{22}+7\,d_{31}+9\,d_{32} -k \leqslant 0, \\ 9\,d_{11}+11d_{12}+13.5d_{21}+16.5d_{22}+9\,d_{31}+11d_{32} -k \geqslant 0, \\ 3.6d_{11}+4.6d_{12}+7.2\,d_{21}+9.2d_{22}+7.2d_{31}+9.2d_{32} \leqslant 4.4, \\ 4.4d_{11}+5.4d_{12}+8.8\,d_{21}+10.8d_{22}+8.8d_{31}+10.8d_{32} \geqslant 3.6, \\ 7\,d_{11}+9\,d_{12}+3.5\,d_{21}+4.5d_{22}+7\,d_{31}+9\,d_{32} \leqslant 5.5, \\ 9\,d_{11}+11d_{12}+4.5\,d_{21}+5.5d_{22}+9\,d_{31}+11d_{32} \geqslant 4.5, \\ d_{11},\ d_{12},\ d_{21},\ d_{22},\ d_{31},\ d_{32} \geqslant 0. \end{cases}$$

The optimal solution of this problem is:

$$k^* = 9.69,\ D^* = \begin{pmatrix} 0.64 & 0 \\ 0.29 & 0 \\ 0 & 0 \end{pmatrix}$$

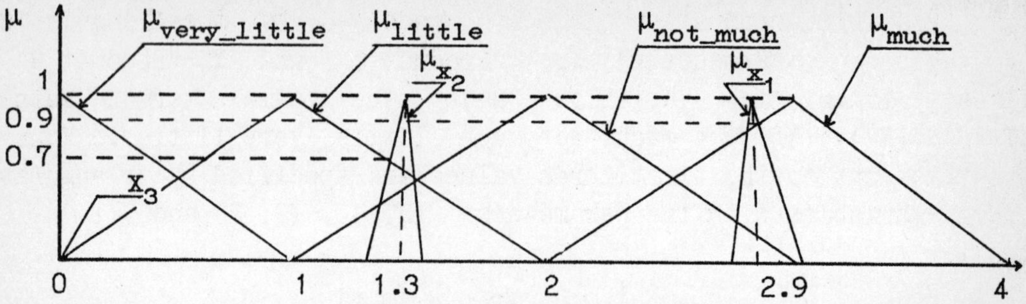

Fig.4. Term set of the linguistic variables "Product_1", "Product_2", "Product_3".

An examplary scenario of a dialog between the user and the system during the input of declarative information may be of the following type :

U Input
S Input_of_the_next_value_of_resource ? (Y/N)
U Y
S Enter_resource_name
U Resource_1
S Enter_resource_value
U Big
S Input_of_the_next_value_of_resource ? (Y/N)
U Y
S Enter_resource_name
U Resource_2
S Enter_resource_value
U Small
S Input_of_the_next_value_of_resource ? (Y/N)
U N

An examplary scenario of a dialog between the user and the system while obtaining the product output plan may be of the following type :

U Plan
S Received_of_the_next_plan ? (Y/N)
U Y
S Enter_product_name
U Product_1
S Output_of_product_1 = much

```
  Possibility_of_plan = 0.9
S Received_of_the_next_plan ? (Y/N)
U Y
S Enter_product_name
U Product_2
S Output_of_product_2 = little
  Possibility_of_plan = 0.7
S Received_of_the_next_plan ? (Y/N)
U Y
S Enter_product_name
U Product_3
S Output_of_product_3 = very_little
  Possibility_of_plan = 1
S Received_of_the_next_plan ? (Y/N)
U N
```

7. CONCLUDING REMARKS

The problem of fuzzy linear programming is considered in this paper. A method of constructing a logical-linguistic rule for this problem - is suggested. It is meant for formalizing the decision making process under uncertainty. Software tools of a decision making support system are represented as an expert system.

This expert system may be classified as a hybride one. It unifies a linear programming package [2] and an expert system which allows to process fuzzy expert knowledge.

The system user has linguistic data. The obtained results are linguistic too. He does not need to understand the system's internal construction. The explanation subsystem is developed to interprete the results obtained.

The possibility distributions are introduced (by the expert and knowledge engineer) on two levels of the expert system:

1) into the optimizations model (2) and (3) to model the uncertainty of the resource vector;

2) to model the user's language, i.e. linguistic variables.

A further development of the system requires the inclusion of some other optimization models.

The expert system is being implemented on the IBM PC - type computer in C under the MS-DOS operating system.

REFERENCES

[1] Averkin A.N., et al. Fuzzy sets in control and artificial intellect models (in Russian)/Edited by Pospelov D.A., Nauka, Moscow, 1986.
[2] Murtagh B.A. Advanced Linear Programming : Computation and Practice. McGraw-Hill, 1981.
[3] Nahmias S. Fuzzy variables, Fuzzy Sets and Systems 1 (1978) 97-110.
[4] Yazenin A.V. A fuzzy mathematical programming (in Russian) University of Kalinin, Kalinin, 1986.
[5] Yazenin A.V. Fuzzy and stochastic programming, Fuzzy Sets and Systems 22 (1987) 171-180.
[6] Yermoliev Yu.M. Stochastic programming methods (in Russian), Nauka, Moscow, 1976.
[7] Zadeh L.A. Fuzzy sets as a basic for a theory of possibility, Fuzzy Sets and Systems 1 (1978) 3-28.
[8] Zadeh L.A. A concept of a linguistic variable and its application to the approximate decision making (in Russian), Mir, Moscow, 1976.

Vol. 261: Th.R. Gulledge, Jr., N.K. Womer, The Economics of Made-to-Order Production. VI, 134 pages. 1986.

Vol. 262: H.U. Buhl, A Neo-Classical Theory of Distribution and Wealth. V, 146 pages. 1986.

Vol. 263: M. Schäfer, Resource Extraction and Market Structure. XI, 154 pages. 1986.

Vol. 264: Models of Economic Dynamics. Proceedings, 1983. Edited by H.F. Sonnenschein. VII, 212 pages. 1986.

Vol. 265: Dynamic Games and Applications in Economics. Edited by T. Başar. IX, 288 pages. 1986.

Vol. 266: Multi-Stage Production Planning and Inventory Control. Edited by S. Axsäter, Ch. Schneeweiss and E. Silver. V, 264 pages. 1986.

Vol. 267: R. Bemelmans, The Capacity Aspect of Inventories. IX, 165 pages. 1986.

Vol. 268: V. Firchau, Information Evaluation in Capital Markets. VII, 103 pages. 1986.

Vol. 269: A. Borglin, H. Keiding, Optimality in Infinite Horizon Economies. VI, 180 pages. 1986.

Vol. 270: Technological Change, Employment and Spatial Dynamics. Proceedings 1985. Edited by P. Nijkamp. VII, 466 pages. 1986.

Vol. 271: C. Hildreth, The Cowles Commission in Chicago, 1939–1955. V, 176 pages. 1986.

Vol. 272: G. Clemenz, Credit Markets with Asymmetric Information. VIII, 212 pages. 1986.

Vol. 273: Large-Scale Modelling and Interactive Decision Analysis. Proceedings, 1985. Edited by G. Fandel, M. Grauer, A. Kurzhanski and A.P. Wierzbicki. VII, 363 pages. 1986.

Vol. 274: W.K. Klein Haneveld, Duality in Stochastic Linear and Dynamic Programming. VII, 295 pages. 1986.

Vol. 275: Competition, Instability, and Nonlinear Cycles. Proceedings, 1985. Edited by W. Semmler. XII, 340 pages. 1986.

Vol. 276: M.R. Baye, D.A. Black, Consumer Behavior, Cost of Living Measures, and the Income Tax. VII, 119 pages. 1986.

Vol. 277: Studies in Austrian Capital Theory, Investment and Time. Edited by M. Faber. VI, 317 pages. 1986.

Vol. 278: W.E. Diewert, The Measurement of the Economic Benefits of Infrastructure Services. V, 202 pages. 1986.

Vol. 279: H.-J. Büttler, G. Frei and B. Schips, Estimation of Disequilibrium Models. VI, 114 pages. 1986.

Vol. 280: H.T. Lau, Combinatorial Heuristic Algorithms with FORTRAN. VII, 126 pages. 1986.

Vol. 281: Ch.-L. Hwang, M.-J. Lin, Group Decision Making under Multiple Criteria. XI, 400 pages. 1987.

Vol. 282: K. Schittkowski, More Test Examples for Nonlinear Programming Codes. V, 261 pages. 1987.

Vol. 283: G. Gabisch, H.-W. Lorenz, Business Cycle Theory. VII, 229 pages. 1987.

Vol. 284: H. Lütkepohl, Forecasting Aggregated Vector ARMA Processes. X, 323 pages. 1987.

Vol. 285: Toward Interactive and Intelligent Decision Support Systems. Volume 1. Proceedings, 1986. Edited by Y. Sawaragi, K. Inoue and H. Nakayama. XII, 445 pages. 1987.

Vol. 286: Toward Interactive and Intelligent Decision Support Systems. Volume 2. Proceedings, 1986. Edited by Y. Sawaragi, K. Inoue and H. Nakayama. XII, 450 pages. 1987.

Vol. 287: Dynamical Systems. Proceedings, 1985. Edited by A.B. Kurzhanski and K. Sigmund. VI, 215 pages. 1987.

Vol. 288: G.D. Rudebusch, The Estimation of Macroeconomic Disequilibrium Models with Regime Classification Information. VII, 128 pages. 1987.

Vol. 289: B.R. Meijboom, Planning in Decentralized Firms. X, 168 pages. 1987.

Vol. 290: D.A. Carlson, A. Haurie, Infinite Horizon Optimal Control. XI, 254 pages. 1987.

Vol. 291: N. Takahashi, Design of Adaptive Organizations. VI, 140 pages. 1987.

Vol. 292: I. Tchijov, L. Tomaszewicz (Eds.), Input-Output Modeling. Proceedings, 1985. VI, 195 pages. 1987.

Vol. 293: D. Batten, J. Casti, B. Johansson (Eds.), Economic Evolution and Structural Adjustment. Proceedings, 1985. VI, 382 pages. 1987.

Vol. 294: J. Jahn, W. Krabs (Eds.), Recent Advances and Historical Development of Vector Optimization. VII, 405 pages. 1987.

Vol. 295: H. Meister, The Purification Problem for Constrained Games with Incomplete Information. X, 127 pages. 1987.

Vol. 296: A. Börsch-Supan, Econometric Analysis of Discrete Choice. VIII, 211 pages. 1987.

Vol. 297: V. Fedorov, H. Läuter (Eds.), Model-Oriented Data Analysis. Proceedings, 1987. VI, 239 pages. 1988.

Vol. 298: S.H. Chew, Q. Zheng, Integral Global Optimization. VII, 179 pages. 1988.

Vol. 299: K. Marti, Descent Directions and Efficient Solutions in Discretely Distributed Stochastic Programs. XIV, 178 pages. 1988.

Vol. 300: U. Derigs, Programming in Networks and Graphs. XI, 315 pages. 1988.

Vol. 301: J. Kacprzyk, M. Roubens (Eds.), Non-Conventional Preference Relations in Decision Making. VII, 155 pages. 1988.

Vol. 302: H.A. Eiselt, G. Pederzoli (Eds.), Advances in Optimization and Control. Proceedings, 1986. VIII, 372 pages. 1988.

Vol. 303: F.X. Diebold, Empirical Modeling of Exchange Rate Dynamics. VII, 143 pages. 1988.

Vol. 304: A. Kurzhanski, K. Neumann, D. Pallaschke (Eds.), Optimization, Parallel Processing and Applications. Proceedings, 1987. VI, 292 pages. 1988.

Vol. 305: G.-J.C.Th. van Schijndel, Dynamic Firm and Investor Behaviour under Progressive Personal Taxation. X, 215 pages. 1988.

Vol. 306: Ch. Klein, A Static Microeconomic Model of Pure Competition. VIII, 139 pages. 1988.

Vol. 307: T.K. Dijkstra (Ed.), On Model Uncertainty and its Statistical Implications. VII, 138 pages. 1988.

Vol. 308: J.R. Daduna, A. Wren (Eds.), Computer-Aided Transit Scheduling. VIII, 339 pages. 1988.

Vol. 309: G. Ricci, K. Velupillai (Eds.), Growth Cycles and Multisectoral Economics: the Goodwin Tradition. III, 126 pages. 1988.

Vol. 310: J. Kacprzyk, M. Fedrizzi (Eds.), Combining Fuzzy Imprecision with Probabilistic Uncertainty in Decision Making. IX, 399 pages. 1988.

Vol. 311: R. Färe, Fundamentals of Production Theory. IX, 163 pages. 1988.

Vol. 312: J. Krishnakumar, Estimation of Simultaneous Equation Models with Error Components Structure. X, 357 pages. 1988.

Vol. 313: W. Jammernegg, Sequential Binary Investment Decisions. VI, 156 pages. 1988.

Vol. 314: R. Tietz, W. Albers, R. Selten (Eds.), Bounded Rational Behavior in Experimental Games and Markets. VI, 368 pages. 1988.

Vol. 315: I. Orishimo, G.J.D. Hewings, P. Nijkamp (Eds.), Information Technology: Social and Spatial Perspectives. Proceedings, 1986. VI, 268 pages. 1988.

Vol. 316: R.L. Basmann, D.J. Slottje, K. Hayes, J.D. Johnson, D.J. Molina, The Generalized Fechner-Thurstone Direct Utility Function and Some of its Uses. VIII, 159 pages. 1988.

Vol. 317: L. Bianco, A. La Bella (Eds.), Freight Transport Planning and Logistics. Proceedings, 1987. X, 568 pages. 1988.

Vol. 318: T. Doup, Simplicial Algorithms on the Simplotope. VIII, 262 pages. 1988.

Vol. 319: D.T. Luc, Theory of Vector Optimization. VIII, 173 pages. 1989.

Vol. 320: D. van der Wijst, Financial Structure in Small Business. VII, 181 pages. 1989.

Vol. 321: M. Di Matteo, R.M. Goodwin, A. Vercelli (Eds.), Technological and Social Factors in Long Term Fluctuations. Proceedings. IX, 442 pages. 1989.

Vol. 322: T. Kollintzas (Ed.), The Rational Expectations Equilibrium Inventory Model. XI, 269 pages. 1989.

Vol. 323: M.B.M. de Koster, Capacity Oriented Analysis and Design of Production Systems. XII, 245 pages. 1989.

Vol. 324: I.M. Bomze, B.M. Pötscher, Game Theoretical Foundations of Evolutionary Stability. VI, 145 pages. 1989.

Vol. 325: P. Ferri, E. Greenberg, The Labor Market and Business Cycle Theories. X, 183 pages. 1989.

Vol. 326: Ch. Sauer, Alternative Theories of Output, Unemployment, and Inflation in Germany: 1960–1985. XIII, 206 pages. 1989.

Vol. 327: M. Tawada, Production Structure and International Trade. V, 132 pages. 1989.

Vol. 328: W. Güth, B. Kalkofen, Unique Solutions for Strategic Games. VII, 200 pages. 1989.

Vol. 329: G. Tillmann, Equity, Incentives, and Taxation. VI, 132 pages. 1989.

Vol. 330: P.M. Kort, Optimal Dynamic Investment Policies of a Value Maximizing Firm. VII, 185 pages. 1989.

Vol. 331: A. Lewandowski, A.P. Wierzbicki (Eds.), Aspiration Based Decision Support Systems. X, 400 pages. 1989.

Vol. 332: T.R. Gulledge, Jr., L.A. Litteral (Eds.), Cost Analysis Applications of Economics and Operations Research. Proceedings. VII, 422 pages. 1989.

Vol. 333: N. Dellaert, Production to Order. VII, 158 pages. 1989.

Vol. 334: H.-W. Lorenz, Nonlinear Dynamical Economics and Chaotic Motion. XI, 248 pages. 1989.

Vol. 335: A.G. Lockett, G. Islei (Eds.), Improving Decision Making in Organisations. Proceedings. IX, 606 pages. 1989.

Vol. 336: T. Puu, Nonlinear Economic Dynamics. VII, 119 pages. 1989.

Vol. 337: A. Lewandowski, I. Stanchev (Eds.), Methodology and Software for Interactive Decision Support. VIII, 309 pages. 1989.

Vol. 338: J.K. Ho, R.P. Sundarraj, DECOMP: an Implementation of Dantzig-Wolfe Decomposition for Linear Programming. VI, 206 pages. 1989.

Vol. 339: J. Terceiro Lomba, Estimation of Dynamic Econometric Models with Errors in Variables. VIII, 116 pages. 1990.

Vol. 340: T. Vasko, R. Ayres, L. Fontvieille (Eds.), Life Cycles and Long Waves. XIV, 293 pages. 1990.

Vol. 341: G.R. Uhlich, Descriptive Theories of Bargaining. IX, 165 pages. 1990.

Vol. 342: K. Okuguchi, F. Szidarovszky, The Theory of Oligopoly with Multi-Product Firms. V, 167 pages. 1990.

Vol. 343: C. Chiarella, The Elements of a Nonlinear Theory of Economic Dynamics. IX, 149 pages. 1990.

Vol. 344: K. Neumann, Stochastic Project Networks. XI, 237 pages. 1990.

Vol. 345: A. Cambini, E. Castagnoli, L. Martein, P. Mazzoleni, S. Schaible (Eds.), Generalized Convexity and Fractional Programming with Economic Applications. Proceedings, 1988. VII, 361 pages. 1990.

Vol. 346: R. von Randow (Ed.), Integer Programming and Related Areas. A Classified Bibliography 1984–1987. XIII, 514 pages. 1990.

Vol. 347: D. Ríos Insua, Sensitivity Analysis in Multi-objective Decision Making. XI, 193 pages. 1990.

Vol. 348: H. Störmer, Binary Functions and their Applications. VIII, 151 pages. 1990.

Vol. 349: G.A. Pfann, Dynamic Modelling of Stochastic Demand for Manufacturing Employment. VI, 158 pages. 1990.

Vol. 350: W.-B. Zhang, Economic Dynamics. X, 232 pages. 1990.

Vol. 351: A. Lewandowski, V. Volkovich (Eds.), Multiobjective Problems of Mathematical Programming. Proceedings, 1988. VII, 315 pages. 1991.

Vol. 352: O. van Hilten, Optimal Firm Behaviour in the Context of Technological Progress and a Business Cycle. XII, 229 pages. 1991.

Vol. 353: G. Ricci (Ed.), Decision Processes in Economics. Proceedings, 1989. III, 209 pages. 1991.

Vol. 354: M. Ivaldi, A Structural Analysis of Expectation Formation. XII, 230 pages. 1991.

Vol. 355: M. Salomon, Deterministic Lotsizing Models for Production Planning. VII, 158 pages. 1991.

Vol. 356: P. Korhonen, A. Lewandowski, J. Wallenius (Eds.), Multiple Criteria Decision Support. Proceedings, 1989. XII, 393 pages. 1991.

Vol. 357: P. Zörnig, Degeneracy Graphs and Simplex Cycling. XV, 194 pages. 1991.

Vol. 358: P. Knottnerus, Linear Models with Correlated Disturbances. VIII, 196 pages. 1991.

Vol. 359: E. de Jong, Exchange Rate Determination and Optimal Economic Policy Under Various Exchange Rate Regimes. VII, 270 pages. 1991.

Vol. 360: P. Stalder, Regime Transitions, Spillovers and Buffer Stocks. VI, 193 pages. 1991.

Vol. 361: C.F. Daganzo, Logistics Systems Analysis. X, 321 pages. 1991.

Vol. 362: F. Gehrels, Essays in Macroeconomics of an Open Economy. VII, 183 pages. 1991.

Vol. 363: C. Puppe, Distorted Probabilities and Choice under Risk. VIII, 100 pages. 1991.

Vol. 364: B. Horvath, Are Policy Variables Exogenous? XII, 162 pages. 1991.

Vol. 365: G.A. Heuer, U. Leopold-Wildburger, Balanced Silverman Games on General Discrete Sets. V, 140 pages. 1991.

Vol. 366: J. Gruber (Ed.), Econometric Decision Models. Proceedings, 1989. VIII, 636 pages. 1991.

Vol. 367: M. Grauer, D.B. Pressmar (Eds.), Parallel Computing and Mathematical Optimization. Proceedings. V, 208 pages. 1991

Vol. 368: M. Fedrizzi, J. Kacprzyk, M. Roubens (Eds.), Interactive Fuzzy Optimization. VII, 216 pages. 1991.